sexual moralities

edited by

83

Clare Hemmings
Irene Gedalof
Lucy Bland

palgrave
macmillan

83 | sexual moralities

contents

sexual moralities
Clare Hemmings, Irene Gedalof and Lucy Bland 1

themed articles

will the real sex slave please stand up?
Julia O'Connell Davidson 4

pole position: migrant British women producing 'selves' through lap dancing work
Esther Bott 23

female sex tourism: a contradiction in terms?
Jacqueline Sanchez Taylor 42

sexual regimes and migration controls: reproducing the Irish nation-state in transnational contexts
Eithne Luibhéid 60

moral ambivalence and irregular practices: contextualizing male-to-male sexualities in Calcutta/India
Paul Boyce 79

revolutionary marriage: on the politics of sexual stories in Naxalbari
Srila Roy 99

making political hay of sex and slavery: Kansas conservatism, feminism and the global regulation of sexual moralities
Almas Sayeed 119

AIDS stigma, sexual moralities and the policing of women and youth in South Africa
Catherine Campbell, Yugi Nair and Sbongile Maimane 132

stakes and states: sexual discourses from New Delhi
Jyoti Puri 139

rewriting 'the paedophile': a feminist reading of *The Woodsman*
Carol-Ann Hooper and Ann Kaloski 149

Sex in development: science, sexuality and morality in global perspective
Carolyn H. Williams 156

The lavender scare: the cold war persecution of gays and lesbians in the federal government
Warren J. Blumenfeld 159

Locked in the family cell: gender, sexuality and political agency in Irish national discourse
Breda Gray 161

Sex, gender and science
Karen Throsby 165

Some of us: Chinese women growing up in the Mao era
Harriet Evans 167

Behind closed doors: domestic violence in India
Kaveri Sharma 169

Women who live evil lives: gender, religion, and the politics of power in colonial Guatemala
Sarah C. Chambers 171

Doing women's studies: employment opportunities, personal impacts and social consequences
Sue Jackson 173

Against empire: feminism, racism and the West
M.S. Sreerekha 175

Feeding anorexia: gender and power at a treatment center
Jo Brain 177

83 | editorial

sexual moralities

At the 25 years of *Feminist Review* celebration event in 2005, Elizabeth Wilson cautiously suggested that for feminist theory to remain relevant for the next 25 years, we might have to relinquish previously treasured objects of study. One of these objects was 'sexuality', the relevance of which to contemporary concerns of globalization, war and poverty was not readily apparent. While not disagreeing with the general point that feminist theory must ever be prepared to interrogate its knowledge base, however difficult or uncomfortable that feels, we are convinced that serious feminist consideration of sexuality remains as important as ever. In a sense this special issue might be understood as responding to the question mark held over sexuality's contemporary political relevance, a counter-assertion of sexuality's central significance within global regimes of power. The pieces included in this issue map the complexities of sexual morality in a range of geographical locations, through a range of disciplinary foci, and through attention to a range of levels of materialization. In relation to the latter, the authors interrogate casual conversations, legal interventions, historical documents, interviews, filmic representations and critical texts, among other arenas, to provide a strong sense of the ways in which sexual morality pervades all aspects of life.

In our original call for submissions, we suggested that a conservative sexual morality appears to be on the increase globally, operating in both new and familiar ways. We were particularly keen to attract work that explored both current and historical manifestations of sexual morality because of the ways in which it is consistently the most vulnerable groups and individuals that are the targets of its logic. Declarations of sexual morality are consistently directed at women and minorities, often as a way of policing national boundaries and preserving the 'integrity' of the heterosexual family. Discourses of sexual morality operate in contradictory ways, frequently serving as a dense site where hierarchies of race, gender, class, and religion are affirmed or challenged. As a result new oppositions emerge – we speak of high- and low-risk groups for sexually transmitted diseases, and sexually liberated versus sexually oppressive nation–states. While cries of sexual morality are increasing, new forms of sexual exploitation and expression reveal themselves – in the global sex trade, in contests between women's rights and foetal rights, in civil partnership legislation for same-sex couples, and in new sexual practices made possible by new forms of communication or mobility.

(1–3) © 2006 Feminist Review. 0141-7789/06 $30 www.feminist-review.com

As this special issue came together, it was clear that many of our concerns were shared by contributors, and that a rich variety of research on sexual morality is currently being undertaken. The majority of articles included here demonstrate the centrality of discourses of sexual morality to the drawing of nation's boundaries. The global circulation of the idea of non-heterosexual, non-familial sexuality as 'Western', for example, operates to keep national citizens of postcolonial nations in line, as both Paul Boyce and Srila Roy illustrate in the context of India. Further, it is through the complex connections among related discourses of nation, sexuality, gender, and class that established hierarchies among subjects within and between nations are managed. Jyoti Puri, Almas Sayeed, Eithne Luibhéid, Campbell *et al.*, Jacqueline Sanchez Taylor, and Esther Bott all unpick the discursive slips and connections that allow sexual regulation to flourish contemporarily, and that position subjects in relation to one another. In each case, this work involves painstaking attention to a local site (New Delhi, Kansas, South Africa, Jamaica and the Dominican Republic, Ireland and Tenerife, respectively) as a way of better understanding the relationship between local and global meanings and flows. This indeed is one of the primary reasons for sustained attention to sexuality: it is frequently at the level of the intimate, the intersubjective or the everyday, as Puri in particular emphasizes, that national and international power relations are produced and sustained.

While we had anticipated a focus in the submissions on new subjects formed through contemporary sexual morality, we were delighted to read work focusing in such detail on the contradictory logics and expectations that sexual subjects negotiate. Boyce's same-sex practitioners, O'Connell Davidson's 'sex slaves' Sanchez Taylor's sex tourist, Bott's migrant lap dancer, Roy's married revolutionary, Hooper and Kaloski's paedophile, and Luibhéid's pregnant asylum seeker are all carefully drawn, real subjects, never 'othered' by their authors, but situated in relation to the local and global contexts that give rise to them. These subjects are never passive, but actively negotiate the conditions of their emergence, in part by constituting others whom they are not. The 'Others' against whom these subjects compare themselves are always more entrenched, less aware, less respectable, more violent, and so on, and for Bott, O'Connell Davidson, and Sanchez Taylor others are consistently racialized or ethnicized. An unsentimental sense of these subjects' precarious agency emerges that, for the authors concerned, is the product of the ambivalences and embeddedness of sexual morality discourse, and evidence of its imbrication in the tangled knots of race, ethnicity, class, gender, and sexual morality.

Reflecting again on the original call for submissions, we had highlighted our interest in articles that examined the role of feminist and women's movements in challenging the pernicious effects of sexual morality. And, in fact, this question of feminism's role in consolidating or challenging sexual morality is a consistent theme across this special issue. Several of the articles point to the limited capacity of current feminist theory or practice to address the complex issues raised by a consideration of sexual morality. Both O'Connell Davidson and Sayeed highlight how, in the UK and US, respectively, a feminist emphasis on legislative intervention to combat the pernicious effects of sexual trafficking can be counter-productive, frequently reinforcing discourses of Western superiority and existing, punitive migration policies. Campbell *et al.* and Boyce challenge feminist and sexuality studies to reflect more consistently on how to translate global HIV and AIDS concerns into local contexts. Without that reflection, they argue, development and policy interventions compound rather than alleviate problems of health-care knowledge and access in South Africa and India. And Hooper and Kaloski and Luibhéid tackle the limitations of some fiercely held feminist presumptions: the paedophile is evil and cannot change, and pro-choice discourses are always in all women's interests. For Hooper and Kaloski, a belief that a paedophile can change challenges the deeply pessimistic assumption that sexual desires and behaviours are set in stone in perpetuity. While for Luibhéid, an analysis of the shifting position on abortion policy in Ireland highlights the ways in which migrant women, and women seeking asylum, may need to have recourse to a foetal rights discourse, however problematic that may be, in order to have any voice at all in legal battles over who has the right to settle within particular national borders.

More tentative, of course, are the suggestions for feminist change that emerge from these reflections, although all the authors are clear that where feminist orthodoxy is inadequate to the task of local–global analysis and action it should be called to account, or indeed abandoned. All authors advocate careful mapping of local contexts as a way of reading the materialization of global sexual discourse, and many suggest the development of flexible translation strategies as an appropriate feminist response to the complexities of sexual morality discourse. In addition, the articles in this special issue all underscore the necessity of feminist approaches attending as centrally to questions of nation, race, ethnicity and class as to gender and sexuality, and demonstrate that a full contextualization of sexual morality is necessary for adequate understanding of its impact. Without this complex attention, it is simply not possible to provide an ethical feminist response to discourses of sexual morality.

Clare Hemmings, Irene Gedalof and Lucy Bland

doi:10.1057/palgrave.fr.9400277

83 will the real sex slave please stand up?

Julia O'Connell Davidson

abstract

This paper critically explores the way in which 'trafficking' has been framed as a problem involving organized criminals and 'sex slaves', noting that this approach obscures both the relationship between migration policy and 'trafficking', and that between prostitution policy and forced labour in the sex sector. Focusing on the UK, it argues that far from representing a step forward in terms of securing rights and protections for those who are subject to exploitative employment relations and poor working conditions in the sex trade, the current policy emphasis on sex slaves and 'victims of trafficking' limits the state's obligations towards them.

keywords

trafficking; sexual slavery; sex work; forced labour; irregular migration

(4–22) © 2006 Feminist Review. 0141-7789/06 $30 www.feminist-review.com

will the real sex slave please stand up?

Over the past decade, a range of governmental and non-governmental actors have displayed growing awareness of, and concern about, the fact that the sex industry can be a site of various (and sometimes extreme) forms of exploitation and abuse. Within this, particular concern has been focused upon phenomena described as 'sexual slavery' and as 'trafficking for sexual exploitation', and in dominant anti-trafficking discourse, two assertions have been so repeatedly made that they have acquired an almost mantra-like quality. The first is that human trafficking is taking place on a massive scale everywhere. Trafficking is described as 'a $7 billion a year business' involving tens of thousands of women and children annually: 'No one now disputes that trafficking today has reached alarming proportions, the magnitude of which affects many countries as countries of origin, transit and destination points' (Javate de Dios, 2002: 1). The second is that trafficked persons are victims of modern slavery and should be treated as such, a statement that is now as likely to be made by government ministers as it is by spokespeople for Non-Governmental Organizations (NGOs) that lobby on the issue.

Given the undisputed and alarming magnitude of the phenomenon, and the fact that for at least the past five years, many international agencies and governments have given high priority and devoted extensive resources to combating the problem, it is puzzling to find that the number of people who have been identified as 'victims of trafficking' (VoTs) and assisted as such is very small. The UK is a case in point. In 2000, a Home Office report estimated that every year, anywhere between 140 and 1400 women and girls were being trafficked into prostitution in Britain, and recommended that the police should pay particular attention to off-street prostitution where trafficked women and girls were especially likely to be held (Kelly and Regan, 2000). New legislation has subsequently been introduced to tackle this supposedly growing phenomenon, a special Home Office funded project (the Poppy Project) has been set up to support VoTs, and various police forces have worked closely with the immigration authorities to identify and rescue women trafficked into the sex industry. However, if those involved in designing and implementing these measures were sincere in their belief that hundreds of women were annually being trafficked into Britain for purposes of sexual exploitation and in their desire to assist them, they must surely be disappointed by the results.

In 2003, the London Metropolitan Police Clubs and Vice Unit and immigration service officers found 295 immigration offenders in the course of regular routine visits to massage parlours and saunas in London, of whom only four or five were identified as VoTs and referred to the Poppy project. The rest were deported (or 'administratively removed', in UK immigration service-speak). In the same year, the Poppy project also received the grand total of 15 referrals from other agencies, and had in all 24 women in their facilities. Raids on indoor prostitution establishments in other areas have yielded similarly low numbers of VoTs. Upon contact with the authorities, the huge and growing numbers of trafficking victims in need of protection have a habit of transmogrifying into 'illegal immigrants' who must be summarily deported. For undocumented migrants in the sex sector, being identified as potentially vulnerable to abuse and exploitation often appears to mean exposure to additional risks rather than an opportunity to access rights and protections. And more generally it seems that despite the extensive attention paid to rights violations in the sex industry and a raft of new international protocols and declarations, national laws, and policy initiatives designed to combat such abuses, very little progress has been made in terms of promoting or protecting the human or labour rights of those who prostitute.

This paper explores the relationship between this lack of progress and the current policy emphasis on sex slaves and trafficking victims. It draws on on-going ESRC-funded research[1] and focuses primarily on the UK, though the same basic argument could be made in relation to policy and practice in many other countries.

1 The study focuses on the markets for migrant domestic and sex workers in the UK and Spain, and includes interview research with those who own or manage sex sector establishments and indirectly employ migrant sex workers, as well as with individuals who buy sex.

slavery, modern slavery and trafficking

To describe human trafficking as 'nothing less than modern day slavery', as UK Solicitor General Harriet Harman recently did (AntiSlavery International, 2005), is a powerful piece of rhetoric. Taken literally, however, it brings us onto difficult philosophical and political terrain, for scholars have always struggled to cleanly demarcate autonomy and slavery, and free and unfree labour, as oppositional categories (Brace, 2004). The problem is that even when understood as defined in the Slavery Convention of the League of Nations (1926), which states that slavery is 'the status or condition of a person over whom any or all of the powers attaching to the right of ownership are exercised', slavery implies a *package* of unfreedoms, not all of which are unique to slavery. Some of the powers attaching to the right of ownership can be and often are also exercised over groups that are not socially imagined as 'slaves', such as wives, children, employees, and professional athletes (Patterson, 1982; Brace, 2004; O'Connell Davidson, 2005). Certainly, if we are concerned with slavery as a form of labour exploitation, there is no clear line between it and 'free' wage labour. Slavery has always stood at one pole of a continuum of exploitation, shading off into servitude and other forms of exploitation, rather than existing as a wholly separate, isolated phenomenon (Lott, 1998).

The quest for a sharp line between slavery and freedom persists in contemporary accounts of 'modern slavery', however. So, for example, in his book *Disposable People: New Slavery in the Global Economy*, Kevin Bales begins by noting that

'New Slavery' is a phenomenon quite separate from other forms of oppression and labour exploitation:

> Having just enough money to get by, receiving wages that barely keep you alive, may be called wage slavery, but it is not slavery. Sharecroppers have a hard life, but they are not slaves. Child labour is terrible, but it is not necessarily slavery.
>
> (Bales, 2000: 5)

New Slavery, according to Bales, also differs from what has traditionally been understood as slavery in a number of senses. It is not a condition that is linked to an excluded legal or political status, it does not necessarily entail natal alienation, it is not a permanent condition, ethnic differences are not relevant to the New Slavery, and the New Slavery is not a feature of a different or separate 'mode of production' – surplus is extracted from the New Slave's labour in essentially the same way that it is extracted from non-Slaves. For Bales, the New Slavery is part of a shadowy, unregulated economic realm in which people can be treated as 'completely disposable tools for making money' (2000: 4). It is the dark and lawless underside of globalization. And because it is impossible to distinguish New Slaves from wage slaves either through reference to their legal or social status, or to the way in which surplus is appropriated by a dominant class, Bales is left insisting that New Slavery differs from wage slavery because it is about 'the *total* control of one person by another for the purpose of economic exploitation' (2000: 6, emphasis added). No matter how hard the life of wage slaves, they are not *completely* controlled by another person. The wage slave can, by implication, make choices.

This notion of 'total control' is a fragile hook on which to hang the concept of slavery. In fact, using this definition, many 'Old Slaves' would not count as 'New Slaves' – not all slaves in slave societies of the Caribbean and America, for instance, were *completely* controlled by their legal owners – some were in a position to engage in a autonomous trading (including the trade of sexual services), and/or to engage in a range of different types of resistance (Beckles, 1989; Lott, 1998; Geary, 2004). And setting up an opposition between 'total control' and 'choice' begs the question of how much choice, and choices between what? If debt – either real or fictional – is used as the means to control a worker, is this complete control or does the debtor have a choice about whether or not to comply with the demands of the person s/he is indebted to? Does an opportunity to quit represent a 'choice' even when it carries with it a risk of being reported to the immigration authorities and deported?

A similar set of philosophical and definitional problems surround the concept of 'trafficking', and boundary disputes are all the more heated because different agencies and groups identify trafficking as a problem for very different reasons and have very different political agendas with regard to the issue. While governments' interest in trafficking is primarily grounded in concerns about irregular immigration and transnational organized crime, human rights NGOs' interest in the phenomenon is often based in broader concerns about 'modern slavery' as defined above. Meanwhile, feminist abolitionist groups such as Coalition Against Trafficking in Women (CATW) view trafficking as central to, and emblematic of, the increasing globalization of female sexual exploitation (Raymond, 2001). Feminist abolitionists hold that it is impossible for women to consent to prostitute, since prostitution dehumanizes and objectifies women. Prostitution is therefore a form of slavery, and since nobody can elect to be a slave, all prostitutes have been trafficked into their condition (Barry, 1995; Jeffreys, 1997).

Until recently, there was no international agreement as to the proper legal definition of trafficking. Following much debate between those with a political stake in the issue, in November 2000, the UN Convention Against Transnational Organized Crime was adopted by the UN General Assembly, and with it two new protocols, one on smuggling of migrants and one on trafficking in persons – the Protocol to Prevent, Suppress and Punish Trafficking in Persons, Especially Women and Children. In the latter, trafficking is defined as:

> The recruitment, transportation, transfer, harbouring or receipt of persons, by means of the threat or use of force or other forms of coercion, of abduction, of fraud, of deception, of the abuse of power or of a position of vulnerability or of the giving or receiving of payments or benefits to achieve the consent of a person having control over another person, for the purpose of exploitation. Exploitation shall include, at a minimum, the exploitation of the prostitution of others or other forms of sexual exploitation, forced labour or services, slavery or practices similar to slavery, servitude or the removal of organs.

The Protocol is held to have redefined the international standard of trafficking in the sense that it allows for the possibility that people may be trafficked for purposes other than exploitation in prostitution, and it is further said to 'establish new standards with respect to protecting the rights of trafficked persons' (Pearson, 2002: 16). The Trafficking Protocol entered into force in December 2003. By 2004, it had been signed by 117 countries, many countries had made efforts to bring their national laws in line with it, and all governments are being urged to do so not only by supranational political bodies such as the Council of Europe, but also by both feminist abolitionist and human rights NGOs involved in anti-trafficking campaigns.

At the same time, however, the Trafficking Protocol has been subject to extensive critique. The Protocol definition of the term 'trafficking' does not describe a single, unitary act leading to one specific outcome, but rather refers to a *process* (recruitment, transportation and control) that can be organized in a variety of ways and involve a range of different actions and outcomes. Trafficking, like traditional understandings of slavery, comes as a package, and there is room for dispute as to which particular actions and outcomes, and in what particular combination, should be included under its umbrella. This problem is compounded by the fact that many of the constituent elements identified in the Protocol definition of trafficking themselves present definitional problems (there is, for example, no international consensus regarding the definition of 'sexual exploitation', or indeed of 'exploitation'), and by the fact that the abuses that come under the umbrella of 'trafficking' can vary in severity, generating a continuum of experience rather than a simple either/or dichotomy (Anderson and O'Connell Davidson, 2002).

Of particular relevance for this paper, the fact that the Protocol is framed within the Convention on Transnational Organized Crime, and packaged with a Protocol on smuggling, reflects a preoccupation with 'illegal immigration' as part and parcel of a supposed security threat posed by transnational organized crime as opposed to a concern with the human rights of migrants (see Beare, 1999; AMC, 2000; Anderson and O'Connell Davidson, 2002; Kapur, 2005). Taken together, the trafficking and smuggling protocols assume a neat line of demarcation between oppositional categories of migration – voluntary and consensual versus involuntary and non-consensual migration – an assumption that research shows to vastly oversimplify the systems and processes that facilitate irregular migration in the real world, and fails to recognize the complexity and variety of social relations between irregular migrants and those who benefit directly or indirectly from their exploitation (see, for instance, Parrenas, 2001; Agustín, 2002; King, 2002; Andrijasevic, 2003; Lutz, 2004). And crucially, the Trafficking Protocol is problematic from a human and migrants' rights perspective because it attaches special significance to situations in which abuses at the point of destination are linked to the use of force or deception within the migration process. State Parties are not being required to meet new and higher standards with respect to protecting the rights of *any* migrant person who is subject to deception, force and exploitation within their borders, but only with respect to those who have also been cheated and exploited within the migratory process.

This latter problem was implicitly recognized in the report of an Experts Group on Trafficking in Human Beings convened by the European Union in 2003, which actually notes that forced labour is the crucial element of the Protocol, and states that 'policy interventions should focus on the forced labour and services, including forced sexual services, slavery and slavery-like outcomes of trafficking – no matter how people arrive in these conditions – rather than (or in addition to) the mechanisms of trafficking itself' (European Commission, 2004: 53). Welcome as this statement is, it does not entirely resolve the problems mentioned above. The concept of 'forced labour' itself presents definitional problems (see Anderson and Rogaly, 2005). Indeed, as the International Labour Office (2005: 8) recent report on forced labour in the global economy acknowledges, 'the line dividing forced labour in the strict legal sense of the term from extremely poor working conditions can at times be very difficult to distinguish'. And even if such definitional problems could be resolved, the Trafficking Protocol would remain a highly selective instrument with which to address the general problem of forced labour because, framed as it is within a Convention on transnational organized crime, the interventions that flow from the Protocol are necessarily only triggered by immigration offences and/or organized criminal activity. As argued below, this means they inevitably focus on an extremely limited and narrow part of the problem.

to catch a victim?

I was recently asked to speak at a conference on trafficking. In the e-mail of invitation, the Conference Producer told me that he was 'very keen to have at least one presentation on the issue of what happens to the victims of trafficking once they have been caught by the authorities'. Doubtless the words were not intentionally chosen, but nevertheless, I felt this mixture of language, the idea of *catching victims*, spoke volumes about the limitations of dominant anti-trafficking discourse from a human rights perspective. Framed as a crime control and prevention issue, 'trafficking' has been presented as a phenomenon orchestrated by ruthless, transnational organized criminal networks. This not only makes it possible for governments to present measures to prevent irregular forms of migration as though they were simultaneously anti-trafficking measures, but also means that the authorities charged with a responsibility to contain illegal migration and combat organized crime are simultaneously deemed to be frontline actors as regards rescuing VoTs. It is implicitly, sometimes explicitly, assumed that by casting a net to catch immigration offenders and individuals involved in a specific set of criminal activities such as those associated with prostitution, drug running, and people smuggling/trafficking, the authorities will also *catch* victims.

Leaving aside, for the moment, the problems and conflicts of interests that may arise when police and immigration officials are expected to catch offenders *and* victims, it is important to note that the waters fished by these actors are extremely limited. The package of violations covered by the UN Protocol definition of 'trafficking' (violence, confinement, coercion, deception and exploitation) can and do occur within legally regulated as well as irregular systems of migration and employment, and within legal as well as illegal systems of migration into private households. (For some extremely disturbing examples of forced labour involving migrant workers who entered the UK through perfectly legal channels and were legally employed in the formal economy, including in the public sector, see Anderson and Rogaly, 2005). And yet migrants who have been subject to abuse and exploitation within legal systems of migration are not likely to be *caught* by police or immigration officials. Indeed, for migrant workers in most sectors in the UK, there is no authority vested with the duty and power to check that they are not being exploited, and if they are, to discover whether their exploitation is linked to the use of force, debt or deception within the migration process.

Although the UN Protocol is often praised for having broadened the international definition of trafficking such that people moved for purposes other than exploitation in prostitution can now be recognized as VoTs, in practice, the sex sector still remains the primary focus of most countries' 'anti-trafficking' efforts. However, identifying VoTs in the sex sector requires the authorities to distinguish between forced labour and extremely poor working conditions, and this is no easy task. Before looking at how police and immigration officials in the UK imagine this distinction, I will briefly illustrate the extent of the problem with examples of employment relations and working conditions in just one form of prostitution in one city, London.

the spectrum of working conditions and employment relations in 'working flats' in London

In November 2005, Gavril Dulghieru, a Moldovan living in London, was convicted for conspiracy to facilitate illegal immigration, misuse of stolen credit cards, forgery, money laundering and conspiracy to traffic for prostitution and sexual exploitation. The women he trafficked worked 20-hour shifts in brothels in Park Lane, Mayfair and Soho, were fed only one meal a day, and charged for the use of cutlery:

> They were forced to have sex with up to 40 men a day for as little as £10 a time to pay off £20,000 debts each – the price for which they were 'bought'. They were charged rent, and subjected to fines if they refused anal or unprotected sex or a client was not attracted to them... One 23-year-old described how she had to pay £300 per day to live locked in a shared basement and her captors threatened to kill her family. Like many trafficking victims, the computer graduate was lured to Britain by promises of a respectable, well-paid job in a hotel or restaurant but ended up in a brothel: 'I believed they would kill my family,' she told the court. 'I thought I hadn't a way out of this situation. I didn't think I had a life in front of me. I wanted to escape but everything was locked. We were locked up all the time. I was told I need to go with clients and I needed to do sex with them. I felt very bad. The first time I wasn't able to talk afterwards'.
>
> (Cowan, 2005)

The 23-year-old quoted above conforms closely to popular understandings and images of a VoT. She was tricked into travelling to the UK with the promise of 'respectable' work, she was locked into a building and forced (both through the use of death threats against her family and through demands to repay a debt) to provide sexual services, and she was psychologically devastated by the experience.

A second example of working conditions and employment relations comes from an interview with a woman, 'Pat', a former sex worker who runs three 'walk up' brothels or flats in Soho, indirectly employing nine sex workers, only one of whom is a British national. Pat does not actively recruit workers either in the UK or abroad, and so has no involvement in facilitating or arranging the migration of the sex workers she employs. Instead, women looking for work (who may or may not be acting under duress from third parties who arranged their migration) telephone or knock on the door and ask to be taken onto her books. The flats are open for business 24 hours per day, every day of the year except Christmas Day and New Year's Day, and the sex workers work 12-hour shifts on a roster system. Only one woman at a time works in each of the flats, which means that on average, every worker on Pat's books takes on between four and five shifts per week, and so works for somewhere between 48 and 60 hours per week. As it is illegal to directly employ a sex worker, Pat's workers do not receive a set wage per shift. Instead, they are required to pay Pat £350 for every shift they work – a sum which in theory pays for the rental of workspace plus other services supplied by Pat (advertising the brothel, employing a maid to take phone calls and clean the flat, provision of food, tea and coffee, also items necessary to their work such as tissues, clean sheets, uniforms, and finally safety precautions such as CCTV).

However, while the relationship between Pat and the sex worker is, in this sense, constructed as a contract between two independent entrepreneurs rather than an employment relation, it is Pat who sets the rates for services and lays down rules concerning working practices, and so, in effect, exerts close control over their work rate. Walk up brothels in Soho cater primarily to the demand for cheap, quick sexual services. Pat's pricing system, which is based on extremely short time units, reflects this. The cheapest service available at £20 is either penetrative sex in the missionary position or oral sex for 10 minutes. If the client wants both penetrative and oral sex in his 10-minute slot, it will cost him £25; if he wants oral sex plus penetrative sex in a position other than missionary, it will cost him £30. The next time unit is 15 minutes, then 20 minutes, then 30 minutes, and then to the highest possible unit, 1 hour, which costs £120 for oral and penetrative sex. Anal sex costs an additional £100, and the prices for oral sex without a condom, and for ejaculating over the worker's breasts or body are also much higher (£140) regardless of the unit of time.

If all the clients that arrived during a shift were to ask for the cheapest service, then in order to pay Pat the shift fee of £350, a worker would need to service 18 clients before even beginning to earn for herself. And if she continued on to service a further 17 clients, she would – like Gavril Dulghieru's forced prostitutes – in effect be having sex with a large number of men for as little as £10 a time. In practice, many clients do ask for more expensive services, and/or can be persuaded to spend more on extras, and sex workers rely in particular upon regular customers who tend to spend longer with them and request more expensive services. Indeed, Pat claims that during a 12-hour shift, workers regularly gross £700, so that they usually take home £350 per day. And yet it is clearly the case that her workers could, on bad days, experience a strong financial pressure to engage in acts that they might otherwise refuse, and that potentially carry serious health risks, such as anal sex, or oral sex without a condom.

As we only interviewed Pat, and not any of her workers, it is not possible to say whether any of the migrants she indirectly employs are working to pay off debts incurred during the process of migration, but other research with migrant women working in the sex industry suggests that it is likely that some are (see Andrijasevic, 2003; Agustín, 2005, for example). The need to repay such debt, especially if combined with the pressure to pay for accommodation in London, and to support dependants back home, could thus operate as strong pressure to continue to work for Pat, also to accept a volume of clients and to provide services that the worker would not otherwise agree to.

A third example comes from an interview with 'Ava', another former sex worker who runs a 'working flat' in the East End of London, and offers indirect employment to 11 sex workers, six of whom are British nationals. The flat is open between 11 am and 10 pm, and each day, two sex workers work the 11-hour shift together. Ava does not, therefore, offer full time employment to the women who work for her (though they may end up working as many hours per week as Pat's workers, since some of them are also indirectly employed in other working flats or massage parlours). Like Pat, Ava does not actively recruit workers – women approach her for work – but as noted in relation to Pat, it is possible that the non-British nationals she employs are working to pay off debts to third parties who arranged their migration. However, if this were the case, the women concerned would not be exposed to the same degree of exploitation as those who work for Pat. Ava provides all the services that Pat provides her workers, but does not charge them a session fee for so doing. Instead, each worker's takings are split equally between Ava and the worker. This means that the worker is not forced to shoulder the cost when trade is slow, and it is not possible for a worker to end up owing money to the house at the end of a session, as it is in establishments that operate the session fee system. Ava sets the prices for services, and they are set at a similar level to those in Pat's Soho walk up. However, unlike at Pat's, there is no 10-minute fee and the price structure and employment relation is not designed to encourage a high throughput of low price trade.

Ava's 'house rules' are also distinctive, leading to different working practices. Ava does not permit her workers to provide any services without condoms, indeed, this rule is important to the way in which she markets her flat in her internet advertising as a clean, safe place to buy sexual services. Furthermore, she actively encourages those who work for her to refuse to provide services that they do not feel comfortable about providing. She explained that she tells workers that it is wrong to assume that they will lose business if they do not agree to anything and everything clients request, saying:

> If anything, it's the other way around, the less you offer, the busier you can be. There's the safety aspect of it, and also the fact that then the guys can look at you and think, 'Well, what she does offer, she must like, because she feels free not to offer what she doesn't like'.

Ava does not have a high turnover of staff. In fact, several of the women who work for her have been with her for a number of years, something that doubtless testifies to the fact that the working conditions and earnings opportunities she provides are good relative to other similar establishments.

will the real sex slave please stand up?

Feminist abolitionist groups like CATW have lobbied very effectively in national and international policy-making circles to get 'sex trafficking' onto the political agenda as a huge and growing problem, and one that involves the enslavement and torture of 'prostituted' women and girls. The campaigning materials they produce reflect their vision of 'the prostitute' as object, victim and slave. They feature graphic depictions of the violence and harm associated with 'the rape-like sex acts of prostitution' (Hughes 2000: 3) – testimony from women and girls who have been violently beaten and raped by pimps, and/or tortured by sadistic clients, lists of physical illnesses and psychological conditions suffered by women in prostitution, images of trafficked women as puppet-like objects or slabs of meat, or even as decapitated heads packaged as sex toys. These images of 'bodies in pain: pierced, bleeding and defenceless' have been reproduced in anti-trafficking materials created by more mainstream organizations, such as the International Organization for Migration and the Organization for Security and Cooperation in Europe, and in government information and awareness campaigns on trafficking (Aradau, 2004: 264).

However, where members of CATW and other feminist abolitionists consider that it is impossible for a woman to consent to sell sexual services, and that *all* migrant women in prostitution (indeed, all female prostitutes, whether migrant or not) are therefore VoTs deserving of protection and assistance, politicians and policy-makers tend to be rather more selective in their view of who is and is not a sex slave. In most countries, to stand any chance of being identified and assisted as a VoT by the authorities, a migrant woman or girl working in the sex trade needs to demonstrate first that she did not choose or consent to work in prostitution, and second that she has undergone great physical suffering. In the US, for example, the Trafficking Victims Protection Act of 2000 makes various forms of protection available to victims of 'severe forms of trafficking', but access to these protections is restricted, and 'relies heavily on the distinction between 'innocent victims' of forced prostitution and 'guilty sex workers' who had foreknowledge of the fact they would be performing sexual labor' (Pearson, 2002; see also Chapkis, 2005: 57–58).

This distinction has much wider currency (Doezema, 1999, 2002; Harrington, 2005). And even where the law does not rule out the possibility that a woman could have agreed to migrate into sex work but subsequently found herself exploited and unable to escape her situation, the status of VoT is by no means automatically given to those who have been subject to the abuses listed as constitutive of 'trafficking' in the UN Protocol. Instead, to be recognized as a VoT, they will need to demonstrate that they have been subject to very specific types of abuse, in particular to rape and/or other forms of extreme physical force. The UK is a case in point. The Asylum and Immigration Bill 2004 states that a person who facilitates travel to or within the UK is guilty of the criminal offence of trafficking if an individual so facilitated: 'is the victim of behaviour that contravenes Article 4 of the Human Rights Convention (slavery and forced labour)', or 'is subjected to force, threats or deception designed to induce him – (i) to provide services of any kind; (ii) to provide another person with benefits of any kind, or; (iii) to enable another person to acquire benefits of any kind'. However, since the legislation, like the UN Trafficking Protocol, fails to provide clear guidelines on the degree of deceit, the type and degree of force, or the type of threats that must be present in order for a person to qualify as a VoT, police and immigration officers must use their judgement to decide whether women and girls picked up during routine visits and raids are or are not VoTs.

Our preliminary interviews with police and immigration officers in the UK suggest that upon contact with the authorities, women need to immediately report a very particular set of experiences to qualify as a VoT. For instance, asked how she could tell if someone picked up during a raid on a massage parlour was or was not a VoT, an immigration officer answered that during their initial interview (which is usually very brief and held in an interview room in a police station), women are questioned about their routing on the way to the UK, whether they knew what kind of work they would be undertaking, whether they had worked in the sex trade back home, and whether an agent was involved in their migration into sex work in UK. And:

> If they say that they were brought here against their will, or they were sold from prostitution in one country on to prostitution in another, or that they're not allowed to leave the building or that they aren't given food, that would obviously tell you they were trafficked.

Likewise, an officer from the Metropolitan Police vice unit explained to me that in his experience, the majority of migrant sex workers in indoor forms of prostitution in London are economically exploited by the owners of such establishments; many have been deceived about earnings and working conditions; and most are working to pay off debts incurred in migration. However, he continued, very few are forced and controlled through means of physical violence or its threat, and it is only rarely that indoor sex establishments are actually run by the same person or people who have recruited the women and/or arranged their entry into the country. In his view, it follows that very few are VoTs. The fact that police and immigration officers in the UK are looking for a very specific constellation of abuses, namely one involving conspiracy to facilitate illegal immigration, plus prostitution forced by means of physical violence or its threat, plus false imprisonment, is further demonstrated by events surrounding a recent raid on a massage parlour in Birmingham. In October 2005, police executed a human trafficking warrant in a massage parlour (named Cuddles) in order to rescue a number of foreign nationals who were believed to be subject to forced prostitution therein. Police and immigration officers removed 19 women from the parlour. Those who could prove that they were legally present in the UK (most of whom were Lithuanian, and so EU citizens) were then released. The conditions under which they were working in Cuddles were not investigated further. The six who could not prove they were legally present in the country were detained and asked, in interviews conducted by male officers and lasting in some cases a mere 17 minutes, how they had travelled to the UK, and how they came to be working in Cuddles. On the basis of their answers, the officers involved decided that they were not VoTs, and two days later, they were transferred to an immigration detention centre to await 'removal'.[2]

2 All six would have been 'removed' four days later had it not been for extensive lobbying with regard to their case. Eventually, the six were interviewed by Poppy Project workers, who identified two of their number as VoTs.

Claudia Aradau has observed that the images of bodies in pain routinely used in anti-trafficking campaigns function:

> 'as a strategy of dis-identification... As trafficked women have been subject to cruelty, their undeniable suffering at the hands of traffickers makes them extra-ordinary, beyond ordinary identifications with illegal migrants and prostitutes. Where their trajectory might have coincided with that of a migrant or prostitute, suffering is redeeming. Trafficked women are dis-identified from categories of migrants, criminals or prostitutes by the emphasis on raw physical suffering'
>
> (Aradau,2004: 257)

Likewise, it seems that physical suffering is the litmus test for police officers and immigration officials involved in sorting VoTs from undocumented migrants working illegally in the sex sector. The threshold of 'victimhood' is, in other words, set very high, and who is going to question this threshold? Not governments that currently place a high priority on being seen to be tough on immigration. And the NGOs that are most vocal in anti-trafficking campaigns are hostages to their own rhetoric. Feminist abolitionists have insisted that violence is an inevitable and ubiquitous feature of the 'sexual slavery' that is prostitution (Weitzer, 2005) and have called on governments to act against 'sex trafficking' on grounds that migrant women and girls in prostitution are routinely subjected to rape, beatings, imprisonment and torture. They are thus hardly in a position to challenge governments to lower the threshold of victimhood. Indeed, the immigration official or police officer who has read nothing about prostitution save the horror stories that are presented as typical by groups such as CATW (stories featuring girls locked in filthy rooms and variously burned with cigarettes, slashed with razor blades, whipped, punched, drugged, etc., etc.) may perhaps be forgiven for assuming that where there is no evidence of this type of physical assault, no rights violations have taken place. Equally, anti-slavery campaigners who define modern slavery as a condition in which a person is *totally* controlled by another person are not well placed to challenge the authorities to work with more complex readings of terms like 'force', 'deception', 'coercion' or 'exploitation'.

the opposite of sex slavery?

Feminist abolitionists are emphatic that all prostitution is sexual slavery. This form of modern slavery has no opposite in the shape of freely chosen sex work. It follows that the victim of sex slavery cannot alter her status as victim and slave by successfully struggling for recognition of her human, civil or labour rights within the sex industry, only by being rescued or freed from her condition as a prostitute. This position has been subject to extensive and powerful critique by other feminist scholars and activists who emphasize women's capacity and right to act as moral agents within prostitution, and who argue that abuse and exploitation in the sex sector is intimately connected to the fact that sex workers are rarely afforded the civil and labour rights enjoyed by other citizens and workers (Alexander, 1997; Bindman, 1997; Chapkis, 1997; Nagle, 1997; Kempadoo and Doezema, 1998; Kempadoo, 1999).

But international policy debate on 'trafficking' and forced labour in the sex sector is not generally informed by a vision of what might constitute the opposite of sex slavery. This is partly because human and child rights NGOs that actively lobby on the issue, as well as many politicians, often assume that the dispute between abolitionists and sex workers' rights activists can simply be left to one side while the real problem is discussed (the idea being that everyone can agree that sexual slavery is wrong, and we should therefore focus on this area of consensus rather than being drawn into an apparently insoluble moral and political debate about prostitution). It is also because religiously inspired and feminist abolitionists together comprise a very economically and politically powerful lobby, one that few can afford to offend. So, for example, the ILO's recent report on forced labour mentioned earlier discusses 'trafficking and forced sexual *exploitation*' and not forced *labour* in the sex sector, and focuses heavily upon the situation of those women and girls who have been deceived into prostitution and on the role of 'powerful organized crime groups' in controlling the sex industry and orchestrating 'human trafficking' (ILO, 2005: 52). Indeed, though the report estimates that 'forced commercial sexual exploitation' represents 11 per cent of cases of forced labour in the contemporary world, and contains much that might be relevant to efforts to secure basic human rights for those working in the most exploitative segments of the sex industry, it does not subject the sex sector to the same kind of analysis as that applied to other sectors. This, it may be assumed, is because the ILO needs US funding to develop its global alliance against forced labour, and such funding would not be forthcoming were the ILO to so much as use the term 'sex work', let alone sketch out a vision of how best to protect sex workers as workers.

Meanwhile, in the UK, the Government's recent review of prostitution (Home Office, 2004, 2005) spoke only to the condition of children, young people, and drug users in prostitution, and those forced to prostitute through the use of physical violence. Other than some recent hints that it may consider measures to allow two or three sex workers to work together in indoor prostitution, the Government has entirely failed to address the situation of those women and men who do regard prostitution as a form of work and whose vulnerability, where it exists, arises precisely because they are denied rights and protections as workers under existing prostitution law, and in the case of undocumented migrants, because of their irregular immigration status.

In short, 'sex slavery' has come to the fore of policy attention, but its opposite in the form of prostitution as work remains largely invisible. In commenting on this, I do not wish to suggest that the rights of those who sell sex can be assured and protected by simply bringing the sex industry above ground and regulating it like any other sector. Indeed, there is no reason to suppose that merely extending existing labour law to cover sex work would automatically safeguard the interests of the most vulnerable workers – see Blackett's (1998: 1) analysis of how, despite their inclusion under general forms of labour regulation in most countries, domestic workers 'remain fundamentally invisible', because most legislative enactments and agreed labour standards do not address the specificity of the employment relationship between domestic worker and employer. And so far as domestic work is concerned, additional problems arise from the fact that employers are often interested in migrants because they are 'less demanding, more flexible concerning working hours'; that migrant domestic workers are often in a 'precarious position because of insecure legal status in the host country'; that for many women, domestic work is often the only way to find employment abroad and to escape poverty in their own country; and that 'unionization of domestic workers is fraught with obstacles' (ILO, 2005: 50–51). These points are also relevant to the sex sector, and highlight very real dilemmas as regards calls for closer state regulation of that sector. For many migrants who trade sex, becoming visible is more likely to mean being deported than it is to mean securing rights and protections as workers.

Some sex workers' rights activists hold that such problems could be solved by granting prostitutes 'the right to travel across state and national boundaries and to obtain work permits on the same basis as other immigrant workers' as well as by making 'sex work amenable to the same kinds of regulations that have reduced harm to workers in… other sometimes hazardous work sites' (Alexander, 1997: 93). However, this overlooks the complex and variable ways in which those who trade sex position themselves in relation to 'prostitution'. At present, much sex commerce takes place in an unregulated underground, or an 'economy of makeshifts' that stands outside civil society understood as 'a social, cultural and ethical system made up of the market, the legal system and voluntary associations to promote the welfare of the community' (Brace, 2002: 334). This is certainly one of the reasons why prostitution is stigmatised, but it is also one of the reasons why those who are excluded from civil society – including undocumented migrants, as well as runaway teenagers, drug addicts, and so on – often turn to it as a means of survival. And it is by no means clear that they would wish to be incorporated into civil society as a 'sex worker', even if this option was open to them.[3]

3 It is also worth noting that some people who prostitute actively wish to live and work in a world outside civil society, untrammelled by state controls and bureaucratic regulations, and entirely free to make individual decisions about the kind of contracts they enter into. Such people may be in favour of decriminalising prostitution, but would not necessarily wish to find themselves fettered by the same kind of regulations that protect (but also constrain) workers in other hazardous work sites.

More generally, there is little cause for confidence in the state as an ally in any struggle to protect the rights of those who trade sex (Chapkis, 1997; O'Connell Davidson, 2003). And yet decriminalization without state regulation, a policy approach advocated by some feminists and prostitute activists, is hardly an attractive alternative, since it endorses 'the sort of minimal regulation on industry that even Milton Friedman would approve' (Shrage, 1994: 83). I do not, then, believe that there is a single, neat and easy policy response to prostitution that will protect all those who trade sex. However, I do believe that policy debate on forced labour in the sex sector cannot be bracketed off from debate on the more general regulation of that sector, any more than the issue of 'trafficking' can be treated in isolation from the more general phenomenon of migration.

In the absence of debate on the specificity of sex work, the details of labour regulation, and the minimum standards that should be applied to the work of prostitutes, and on what constitutes an unacceptable level of 'exploitation' within prostitution (self-exploitation as well as by an employer), it is dangerous to talk about forced prostitution and sex slaves. For without some agreement on the standards and norms that *should* apply, it is impossible to say which side of the line separating forced labour from extremely poor working conditions women like those working for Pat fall on, and therefore impossible either to take steps to assist them as victims of forced labour, or to identify strategies for improving their situation as 'free' workers undertaking 'poor work'. If we cannot, for example, identify a maximum number of clients per day that a sex worker can be expected to service, a minimum payment per sexual service provided, a maximum number of times that a sex worker can safely agree to provide anal sex, and minimum standards with regard to other health and safety issues specifically associated with sex work, then there is no way of distinguishing between the three employers described earlier except through reference to the use of physical force.

In the UK, the current immigration regime and the absence of labour standards in the sex sector combine to leave many irregular migrants in a position whereby they have little choice but to accept extremely poor working conditions and highly exploitative employment relations. They can expect (in theory if not in practice) nothing more from the state than to be protected from an employer who locks them into a building, and/or rapes or beats them or threatens to kill their family. Such a very low standard of nothingness effectively equips employers with a *carte blanche* to make whatever contracts they please with sex workers, and perhaps the most remarkable thing about the sex industry is the fact that there are actually some employers, like Ava, who attempt to raise the standard beyond the extraordinarily basic minimum requirement that they should not use physical force against their employees. Far from representing a step forward in terms of securing rights and protections for those who are subject to exploitative employment relations and poor working conditions in the trade sex, the current policy emphasis on sex slaves and VoTs limits the state's obligations towards them.

acknowledgements

I am grateful to the Economic and Social Research Council, which funded the research on which this paper is based (Award No. R000239794), and to Bridget Anderson, who co-holds this award and with whom the ideas presented in this paper have been developed.

author biography

Julia O'Connell Davidson is Professor of Sociology at the University of Nottingham. She has been involved in research on various aspects of the commercial sex trade since 1993, and is currently working on an ESRC-funded project exploring the markets for migrant sex and domestic workers in the UK and Spain. She is the author of *Prostitution, Power & Freedom* (Polity, 1998) and *Children in the Global Sex Trade* (Polity, 2005).

references

Agustín, L. (2002) 'Challenging 'place': leaving home for sex' *Development*, Vol. 45, No. 1: 110–116.

Agustín, L. (2005) 'Migrants in the Mistress's house: other voices in the 'trafficking' debate' *Social Politics*, Vol. 12, No. 1: 96–117.

Alexander, P. (1997) 'Feminism, sex workers and human rights' in J. Nagel, editor, *Whores and Other Feminists*, London: Routledge.

AMC (2000) *Asian Migrant Yearbook 2000: Migration Facts, Analysis and Issues in 1999*, Hong Kong: Asian Migrant Centre.

Anderson, B. and O'Connell Davidson, J. (2002) *Trafficking – A Demand Led Problem? Part I*, Stockholm: Save the Children Sweden.

Anderson, B. and Rogaly, B. (2005) *Free Market, Forced Labour?*, London: TUC.

Andrijasevic, R. (2003) 'The difference borders make: (il)legality, migration and 'trafficking' in Italy among 'eastern' European women in prostitution' in S. Ahmed, C. Castaneda, A. Fortier and M. Sheller, editors, *Uprootings/Regroundings: Questions of Home and Migration*, Berg.

AntiSlavery International (2005) *Trafficking Network Update*, January. London: ASI.

Aradau, C. (2004) 'The perverse politics of four-letter words: risk and pity in the securitisation of human trafficking' *Journal of International Studies*, Vol. 33, No. 2: 251–277.

Bales, K. (2000) *Disposable People: New Slavery in the Global Economy*, Berkeley, CA: University of California Press.

Barry, K. (1995) *The Prostitution of Sexuality*, NY: New York University Press.

Beare, M. (1999) 'Illegal migration: personal tragedies, social problems, or national security threats?', in P. Williams, editor, *Illegal Immigration and Commercial Sex: The New Slave Trade*, London: Frank Cass.

Beckles, H. (1989) *Natural Rebels: A Social History of Enslaved Black Women in Barbados*, Bloomington: Indiana University Press.

Bindman, J. (1997) *Redefining Prostitution as Sex Work on the International Agenda*, London: Anti-Slavery International.

Blackett, A. (1998) *Making Domestic Work Visible: The Case for Specific Regulation*, Geneva: International Labour Office.

Brace, L. (2002) 'The tragedy of the freelance hustler: Hegel, gender and civil society' *Contemporary Political Theory*, Vol. 1: 329–347.

Brace, L. (2004) *The Politics of Property: Freedom and Belonging*, Edinburgh: Edinburgh University Press.

Chapkis, W. (1997) *Live Sex Acts*, London: Cassell.

Chapkis, W. (2005) 'Soft glove, punishing fist: the Trafficking Victims Protection Act of 2000' in: E. Bernstein and L. Schaffner, editors, *Regulating Sex*, London: Routledge.

Cowan, R. (2005) 'Immigrants tell of forced prostitution and slavery as trafficking gang is jailed' *The Guardian*, November 2.

Doezema, J. (1999) 'Loose women or lost women? The re-emergence of the myth of 'white slavery' in contemporary discourses of trafficking in women' *Gender Issues*, Vol. 18, No. 1: 23–50.

Doezema, J. (2002) 'Who gets to choose? Coercion, consent, and the UN Trafficking Protocol' in R. Masika, editor, *Gender, Trafficking and Slavery*, Oxford: Oxfam.

European Commission (2004) *Report of the Experts Group on Trafficking in Human Beings*, Brussels: European Commission.

Geary, D. (2004) 'Europe and slave protests in the Americas (1780–1850)' *Mitteilungsblatt des Instituts fuer Soziale Bewegungen*, Heft 31. Bochum: Ruhr University.

Harrington, C. (2005) 'The politics of rescue: Peacekeeping and anti-trafficking programs in Bosnia-Herzegovina and Kosovo' *International Feminist Journal of Politics*, Vol. 7, No. 2: 175–206.

Home Office (2004) *Paying the Price: A consultation paper on prostitution*, London: Home Office.

Home Office (2005) *A Coordinated Prostitution Strategy and a Summary of Responses to Paying the Price*, London: Home Office.

Hughes, D. (2000) 'Men create the demand, women are the supply' *Feminista!*, Vol. 4, No. 3.

ILO (2005) *A Global Alliance Against Forced Labour*, Geneva: International Labour Office.

Javate de Dios, A. (2002) 'Revisiting the issue of trafficking in women: Comments on policy implications of a gender and rights framework. Paper presented to ASEM Seminar, Promoting

Gender Equality to Combat Trafficking in Women and Children, organised by the Swedish Ministry for Foreign Affairs. Bangkok, 7–9 October.

Jeffreys, S. (1997) *The Idea of Prostitution*, Melbourne: Spinifex.

Kapur, R. (2005) *Erotic Justice: Law and The New Politics of Postcolonialism*, London: Glasshouse.

Kelly, L. and Regan, L. (2000) 'Stopping traffic: Exploring the extent of and responses to trafficking in women for sexual exploitation in the UK' *Police Research Series*, Paper 125. London: Home Office.

Kempadoo, K. (1999) 'Slavery or work? Reconceptualising Third World prostitution' *Positions*, Vol. 7, No. 1: 225–237.

Kempadoo, K. and Doezema, J. (1998) *Global sex workers: Rights, Resistance and Redefinition*, New York: Routledge.

King, R. (2002) 'Towards a new map of European migration' *International Journal of Population Geography*, Vol. 8: 89–106.

Lim, L. (1998) *The Sex Sector: The Economic and Social Bases of Prostitution in Southeast Asia*, Geneva: International Labour Office.

Lott, T. (1998) 'Early Enlightenment conceptions of the rights of slaves' in T. Lott, editor, *Subjugation and Bondage*, Rowman and Littlefield.

Lutz, H. (2004) 'Life in the twilight zone: migration, transnationality and gender in the private household' *Journal of Contemporary European Studies*, Vol. 12, No. 1: 47–55.

Nagle, J. (1997) *Whores and Other Feminists*, London: Routledge.

O'Connell Davidson, J. (2003) 'Sleeping with the enemy? Some problems with feminist abolitionist calls to penalise those who buy commercial sex' *Social Policy & Society*, Vol. 2, No. 1: 55–63.

O'Connell Davidson, J. (2005) *Children in the Global Sex Trade*, Cambridge: Polity.

Parrenas, R. (2001) *Servants of Globalization: Women, Migration and Domestic Work*, Stanford, CA: Stanford University Press.

Patterson, O. (1982) *Slavery and Social Death*, Cambridge, MA: Harvard University Press.

Pearson, E. (2002) *Human Traffic, Human Rights: Redefining victim protection*, London: AntiSlavery International.

Raymond, J. (2001) *Guide to the New UN Trafficking Protocol*, North Amherst, MA: CATW.

Shrage, L. (1994) *Moral Dilemmas of Feminism*, London: Routledge.

Weitzer, R. (2005) 'Flawed theory and method in studies of prostitution' *Violence Against Women*, Vol. 11, No. 7: 934–949.

doi:10.1057/palgrave.fr.9400278

83 pole position: migrant British women producing 'selves' through lap dancing work

Esther Bott

abstract

This paper explores the motivations and experiences of British women working as lap dancers in the tourist resorts of southern Tenerife, with a particular focus on the subjective choices and processes undertaken by working-class women in the embodiment of positively evaluated identities. It uses Skeggs' theoretical framework of 'becoming respectable' (1997) alongside other debates on 'identity management' in order to begin mapping the ways in which migrant British lap dancers produce themselves, negotiate gender and class, and seek forms of respectability, reputability and honour through their work. Drawing on empirical data, it will discuss how strong disassociations with the Other are formed, and how and why important binaries, particularly distinctions between lap dancing and prostitution and lap dancing and other 'degraded' forms of work and lifestyle are drawn in the dancers' own stories of themselves. It will look at notions and processes of gaining 'respectability' through part of a wider migrant discourse, exploring how being a migrant lap dancer can serve as a vehicle for reproducing a 'respectable' and otherwise positively evaluated self on several levels.

keywords

lap dancing; identity management; distinction; British migration; embodiment; sex work boundaries

(23–41) © 2006 Feminist Review. 0141-7789/06 $30 www.feminist-review.com

introduction

In her highly influential and important *Formations of Class and Gender* (1997), Beverly Skeggs generated questions about the theoretical frameworks used to understand how women's subjectivities are constructed, identifying weaknesses in the class/gender linkages within analyses of gender construction and reproduction. Skeggs explored how women negotiate and understand their classed and gendered 'selves' through social and cultural relations. More specifically, the study exposed how white working-class women can access notions of respectability through occupational, social and cultural identification. Skeggs reminds us that respectability is primarily the concern of those to whom it is not automatically granted, and that those in pursuit of it do so in order to cast off the dirt and danger historically associated with the working classes. Her longitudinal ethnographic study, which traces the life courses of women over a 12-year period, begins at the point of their enrolment on a Social Care course at a Further Education college, and follows their journey through the labour market and into family life.

In mapping the significance of respectability to subjective class categorization, Skeggs' research explored the notion of class *dis*identification. She discovered that the women experienced class not through explicit identification with class positioning but rather by a deliberation to *not* be working-class: '[T]hey made enormous efforts to distance themselves from the working-class' (1997: 74). But processes of class *dis*identification did not manifest directly either. Instead, they happened in what Skeggs calls a 'dialogic manner' of comparison: 'in every judgement of themselves a measurement was made against others' (1997: 74). Skeggs discusses processes of 'valorization' as occurring by disassociation with an unvalorized social identity, that is, working-class woman, and the accessing of new, positively evaluated identities as always being set in opposition to the Other.

It is my aim in this paper to discuss similar processes of disidentification with a disreputable Other and the idea of comparative subjectivities in another context. My own work on British migration to Tenerife involved ethnographic research into lap dancing, a sector of the tourism industry in Tenerife and other destinations which is undergoing growth and rising popularity among young British women migrants.

A limited literature on 'exotic dancers' in the USA has been exploring the interactions and relationships between dancers and customers (Enck and Preston, 1988; Frank, 1998; Wood, 2000). It has also begun mapping and theorizing entry into exotic dancing (Lewis, 1998; Forsyth and Deshotels, 1997), and has examined the ritualization of demand from regular customers (Frank, 2002). However, very little attention has been dedicated to understanding the social positioning and subjectivities of the dancers, or the meanings of lap dancing as a form of work, or the discursive and/or definitional distinctions between exotic/lap dancing and prostitution. The growing popularity of lap dance and other 'exotic' dance venues in a British context also remains under-researched, as does the rising phenomenon of this particular sex industry sector in holiday and migration destinations popular with the British. This paper aims to capture the British migrant lap dancing scene in Tenerife, and to use Skeggs' theoretical framework of 'becoming respectable' alongside other debates on 'identity management' in order to begin mapping the ways in which migrant British lap dancers produce themselves, negotiate gender and class, and seek forms of respectability and reputability through their work and consumption. It will discuss how important distinctions between lap dancing and prostitution and lap dancing and other 'degraded' forms of work are drawn in the dancers' own stories of themselves. It will look at notions and processes of gaining 'respectability' through part of a wider migrant discourse, exploring how being a migrant lap dancer can serve as a vehicle for reproducing a 'respectable' and otherwise positively evaluated self on several levels.

methods

The paper draws on wider data gathered in the course of ongoing ESRC-funded research focusing primarily on the markets for migrant domestic and sex workers in Spain and the UK, but which also involves ethnographic research on the informal tourism economy (and its intersections with the domestic and sex work markets) in Tenerife. The data stem from 16 semi-structured interviews with British dancers from three different lap dance venues, as well as additional interview and survey data from a wider sample of third parties (club owners, club DJs) and clients. All informants were British people living in Tenerife.

The methodological and ethical concerns of the study (including gaining access, negotiating researcher/researched relations and power dynamics, sensitivity and reflexivity in the field, consent and confidentiality, researcher safety and so on) are important and complex, and a serious discussion of them is beyond the capacities and focus of this paper. Here I have room to mention only some key points: three lap dance clubs in close proximity to each other were selected as they were reputed to employ mainly British women and were located in the main tourist 'strip' popular with British tourists. Gaining access to the dancers presented difficulties and usually required me to negotiate with a gatekeeper (club manager or sometimes doorman or DJ). Once I had obtained gatekeepers' permission to approach the women, persuading them to talk to me was often tricky, especially at first. They tended to be wary of me for a number of reasons that I could identify. Some suspected that I was a tabloid journalist working undercover; others were concerned that I might be there to make moral judgements ('so what do you want to know – about how *bad* we all are?'). Some were simply irritated by my presence and I often felt awkward and interfering. On further visits, after the first couple of interviews had taken place, however, the women became less inclined to ignore me and several even approached me to give an interview.

Some interviews were carried out inside clubs, usually seated at the bar, between 9pm and approximately 10.30pm, after which time the dancers generally became too busy to talk. I was able to arrange several follow-up interviews, which were conducted during the day and in a neutral location. These follow-up interviews proved to be more revealing, partly because I was able to record them but also perhaps because the interviewees felt more relaxed being fully clothed, away from the demands and restrictions of work and in a quieter environment. Of all the women I asked for an interview, only one refused but then later agreed to meet me elsewhere at a later date. One woman who gave me a brief, interrupted interview in a club agreed to do a follow-up interview for which she did not turn up.

This study, which is above all interested in the (re)production of 'selves' is not without limitations. Unlike Skeggs' extensive, long-term study on processes of 'becoming respectable', this research offers merely a snapshot of a particular time in these women's lives. The study does not lend itself to a longitudinal repeat and cannot make claims about the longer-term trajectories of its subjects. While providing an important insight into the current subjective experiences of some women in an under-researched area, it should be acknowledged that the 'selves' being produced have potentially only a short shelf-life (women rarely continue lap dancing much beyond their early thirties) and it would be nearly impossible to study the interviewees at a later stage in their lives in order to explore whether the capitals and self-identities they developed through lap dancing are enduring or transferable through time or into other working environments.

migration to Tenerife

Similar limitations may also apply to understandings built upon analyses of the often very short-term migration of British people to Tenerife. Elsewhere (Bott, 2004) I have discussed British migration to Tenerife as being illustrative of some of the new, unorthodox migration motives concerning contemporary writers on migration (Cohen, 1995; King, 2002; Castles and Miller, 2003). King discusses 'changing views of consumption and self-realization' as 'important implications for defining and studying migration' (2002: 90). Young British migrants to Tenerife can be discussed within a discourse of such incentives, with ideas around escape, self-improvement and status seeking – in other words, ideas connected to identity reproduction/management, often generated through 'valorized' forms of work – being key identifiable migration motives (Bott, 2004).

Among the employment options open to young British women who have, in some form or another, migrated to Tenerife, lap dancing is one of the better paid. Of all 16 lap dancers whom I interviewed, only one had had lap dancing work in mind when she migrated to Tenerife from the UK. The others had drifted into it from other forms of service work, usually serving drinks or touting on behalf of bars or lap dance clubs. Lap dancing brings significantly higher earning potential than do bar work and touting. Bar work is the most reliable source of income but is poorly paid – usually around €12–20 per shift plus a drinks allowance. PR/touting is paid by commission on customers brought in, or by a combination of a low basic wage plus commission, usually €1 per head for diners or 50 Eurocents for those just wanting drinks. A small number of women work in holiday timeshare sales, which though potentially well paid, can also be extremely hazardous and competitive, and is very male dominated.

lap dancing

Lap dancing is a form of 'exotic dance'. An exotic dancer has been defined as 'one who removes all or most of her clothing in a sexually suggestive fashion to a paying audience in a performance environment' (Bernard *et al.*, 2003: 2). Other forms of exotic dance include 'striptease', 'table dancing' and 'pole dancing'. Lap dancing specifically involves a form of one-to-one interaction between dancer and a paying customer in his or her 'lap' area, although many clubs operate strict 'no touch' rules. The lap dance itself is a 'personal' performance in very close proximity to the paying customer but which may be viewed by other members of the audience. Some clubs offer private booths, where this interaction can take place in relative isolation from the rest of the club, although there is often a gap of half a meter or so between the booth door and the floor, through which the feet and legs of the dancer and her client are visible, creating a peep-show type environment. Several clubs in Tenerife encourage or demand their dancers to perform 'extras' (sex or sexual services involving physical contact for extra money) in such booths. Lap dance venues often have free entry (although drinks sell at inflated prices) and club owners take a cut from dances sold. In Tenerife, dances cost between €12 and €20 and last the length of one song (approximately 3–6 minutes). Club owners receive roughly 25% from the price of each dance.

In the clubs visited during these studies (which I will call clubs 'X', 'Y' and 'Z') between five and 12 women were working every night. Each dancer works six nights per week from 10pm until between 3am and 5am, depending on how busy the club is. The majority of dancers were British women, although there were

Hungarian, Bulgarian, Czech, Russian and South American dancers also. The dancers arrive at work for 9pm and then spend about an hour getting ready to perform, applying make-up, rehearsing, chatting, and drinking soft drinks from the bar. At around 10pm, customers begin to arrive at the club and the DJ turns up the volume. In one particular venue at around this time each night, the manager filled a paper bag with slips of paper displaying a different number (1–20) and each dancer picked a number at random from the bag. The number denotes the order in which the girls must perform a striptease (to naked) and a pole dance, again for the duration of one record. The women continue taking to the stage in the same order throughout the night.

Members of the audience are thus able to decide which dancer(s) they would like to buy a lap dance from. They normally have to negotiate and pay for dances at the bar. Dances can be paid for singularly or in multiples, with cash or by credit card. Usually, and especially during quiet times or when venues are overstaffed with dancers and competition is high, the dancers must engage in lengthy, flirtatious conversations with potential customers in order to sell dances.

The dancers ranged from age 17 to 32 years, the majority being in their late teens or early 20s. Several of the women interviewed said they had been working as lap dancers since they were 16 years old and had lied about their age to employers. All of the women were white and described their backgrounds as 'working-class'.[1] Like many of the young British migrants interviewed and surveyed in the wider study of British migration to Tenerife, almost all the dancers said that they had arrived in Tenerife with very little money, no employment and no accommodation. Many travel to Tenerife on cheap package tours and deliberately miss the return flight home.

With regard to demand, interviewees held that their clients include many male British regulars, especially timeshare salesmen and other tourism sector workers, who are often 'coked-up' but 'bearable'. Lap dance clubs in Tenerife also attract tourists from Britain, Germany, Scandinavia and Italy as well as Latin American regulars who reside in Tenerife. They also appeal to a small number of women tourists, mainly from Britain, German and Scandinavia.

British dystopia

Boredom, mundanity, poor living standards and lack of opportunity in the UK emerged as central, explicitly stated reasons for young British lap dancers (and indeed young British migrants in general) wanting to leave the UK (Bott, 2004). Living and doing lap dancing in Tenerife can serve as vehicles for renouncing a negatively evaluated British working-class background on several levels. First, and most obviously, is that they constitute a geographical departure from one's place of birth. This in itself is a statement of mobility, of independence, of

1 I acknowledge the definitional and political difficulties surrounding the terminology of class. The wider study into British migration to Tenerife, to which this research on lap dancing contributes, was concerned with identity and 'mobility' narratives. Interview schedules therefore included the question 'How would you describe your class background in the UK?'. All 16 women interviewed here answered 'working-class' or a variation thereof, for example, 'ordinary working-class people'.

freedom (albeit politically structured), of bravery, of doing something, which, I was told by an interviewee, her parents wish they had done years ago. Elsewhere (Bott, 2004) I have discussed in detail the emergent motives behind British migration to Tenerife, and a central theme is that Tenerife offers an opportunity for salvation. Britain is widely perceived as being a place where 'ordinary people' become trapped in a life of misery. A clear manifestation of this idea emerged in more than half of the total number of interviews conducted, where interviewees held onto a secondary subjectivity – their unmigrated self – and described to me the alternative story of their lives, the one they would be living had they not had the resourcefulness to escape to Tenerife.

For example, Karen had worked in clubs X and Z alternately for 9 months. She had just turned 18 when I interviewed her and was from a large town in the north of England. She had left school with 3 GCSEs and had worked in retail for several months, earning below the minimum wage and living at home with her parents. Her father's continued unemployment and the breakdown of her parents' marriage caused her relationship with them to deteriorate. Just before her 17th birthday, Karen ran away to Tenerife with a friend. They had purchased a two-week package holiday with self-catering accommodation included and had planned to skip the flight home and rent an apartment together in Playa de las Americas. At the end of the two-week period, however, the friend changed her mind and went home to the UK. Karen felt that she could not return home and had very little money left. She explained what she did next:

> I slept on the beach for a few nights, probably about a week, on the sun beds, which is where all the shaggers go so it was fairly horrible. Then I met some girls and they let me sleep on their settee till I found a job. I did some PRing for a bit, but the money was so crap, I couldn't afford my own place on that money. Then one of the girls I was staying with said why don't you try lap dancing, 'cause the money's good. I thought 'no way' but I tried it and I thought I was really crap at it and I felt really ashamed. The first night I went home and cried. I was so nervous and I thought there was no way I could do it. But I stuck at it and got better. Now I don't mind it, I'm quite good and I earn enough money to rent a flat with my boyfriend. I usually make about a hundred [Euros] a night doing a few hours exercise. Can't be bad can it?

I asked Karen if she anticipated returning home to the UK at any point. She replied:

> No way. I mean, if I had a degree like you, I'd be fine wouldn't I? There's absolutely nothing for me to go back to. I've got no qualifications so all I'd end up doing is working in Dorothy Perkins for five quid an hour being told what to do all day long, making money for someone else. I can't even stand to think about that. When I think about going back all I see is this person who is pale and skint and hanging about with a load of single-mothers. I'd end up killing myself.

Karen had a very clear picture of her 'alter' self. This kind of 'alterity' narrative appears to serve as an important prompt or incentive to persevere with what can be very demanding and demoralizing work. Karen compares herself favourably with an alternative, 'pale and skint' self who has no choice but to work in 'mundane' retail jobs and become enormously depressed. But there is a second Other in Karen's narrative; the working-class women she left behind in her hometown whose inevitable fate involves getting pregnant and becoming single-mothers. Karen sees this as the only real alternative for women of her background, and part of being *her* is about not being *them*. Skeggs (1997) discusses this as a process of disidentification with the limited and negative connotations of being a working-class woman. Karen's own negative reference to such a pathologizing representation of working-class women (as all being single-parents) demonstrates her own disassociation with that particular version of working-class femininity, and exemplifies the generation of what Skeggs (1997: 76), as well as Steedman (1986), Walkerdine and Lucey (1989) and Walkerdine (1990), have noted as a longing to be in a *different place*.

Part of realizing such a longing comes from migrating out of an unfavourable, undesirable situation to a warm, sunny, 'exotic' location; from turning away from one's history and by not being that pale, failed, Britain-bounded self. But what part does becoming a lap dancer contribute to arriving at a 'different place'?

A good starting point is to consider that lap dancing can be extremely physically, emotionally and psychologically demanding work. Overcoming these difficulties can in itself bring, perhaps not respectability, but certainly a sense of honour. Pitching herself against the difficulties and dangers associated with her work mobilizes a sense of pride and feelings of achievement for Laura, an 18-year-old from the East Midlands. The adversities of lap dancing and the subsequent honour she experiences in overcoming them are compounded and intensified by the migration context of her work:

> [I]t's hard and you need balls to do it, which I didn't think I had and it makes me proud that I can stand up there and strip and get on the pole. I'm actually good at it and that makes me proud. It's not easy, you know, coming to a foreign country where you don't know many people and getting up there on the stage and stripping off. I've never been good at anything in my life before.

Managing to be a migrant and a lap dancer, and, moreover, a *good* lap dancer is important to Laura's positively valued subjectivity. Lap dancing in Tenerife is viewed as something that takes foresight and considerable tenacity to achieve, and it represents a world far removed from the drabness and tedium of council estate life and single-parenthood. Lap dancing, despite its difficulties, remains glamorous:

> I was a bit ashamed at first to strip naked but I soon got used to it. Now I just really love the feeling I get from dancing. It's about the only time I feel strong and sexy; I don't think it's degrading at all. It gives me a massive buzz. I even get on the pole in clubs when I'm out on my night off. It's given me loads and loads of confidence. There's something about being up there on that stage all dressed up

> with your hair done and everybody's looking at you. You have to keep yourself looking good, your nails, your skin, and you always have to have nice underwear and dresses. I love all that. And I love how slim and fit my body is – I've even got a six-pack!

This was Rachel's response to a question about what she likes and dislikes about lap dancing. Rachel is 19 years old and comes from Yorkshire. She had never done lap dancing before she came to Tenerife one year ago. She had previously worked in a factory in the UK, which she hated and described to me as 'really heavy, boring work' and the factory as being 'full of really bitchy old hags'. Rachel experiences lap dancing as a positive means for being glamorous and for expressing a glamorized femininity, which is projected against the backdrop of the factory. Like Karen, Rachel sets her self against a negative, pathologizing representation of working-class women, her ex-colleagues at the factory, whom she describes as old and 'bitchy'. Their haggishness contrasts sharply with her glamour, vigour, beauty and youth (see Table 1(1)).

The women of Skeggs' study were claiming a caring, responsible respectability through their disidentification with a dirty, dangerous, valueless Other. A caring, responsible respectability can be chased institutionally, through marriage and the family, and gratification can be deferred, through college courses and training which lead to respectable caring jobs. Skeggs notes that the women she researched consolidated themselves as valued and caring by performing caringly and altruistically, and that this moral superiority enabled a location with respectability. The nurturing, caring representation of femininity could be embodied and the dirty, dangerous Other disembodied. In tilted reflection of this process, the women of this study can be said to be finding access to a reputable, recognizable and, in and through their own value placements and disidentifications, *respectable* femininity through their dissimulation with ordinariness, sexlessness and failure to escape. The young British lap dancers can embody another version of idealized femininity, that of the long-haired, made-up, slim and attractive, recently waxed and bronzed beauty.

Lap dancing has become more popular, more professionalized and less underground in the UK in recent years. The 'high class' image of lap dance club chains such as *Spearmint Rhino,* the appearance of supermodel Kate Moss performing a semi-naked pole dance in a recent *White Stripes* video and news stories about middle-class university students supplementing their incomes with lap dancing work have done a lot for the image of lap dancing. The embodiment of the glamorous and corporeally 'ideal' lap dancer can mobilize the rejection of her opposite number – pale, haggish, trapped – in the UK.

If, as Skeggs did, we look to Bourdieu to frame an analysis of the relationship between becoming respectable and embodiment, we can discuss the body as a site for improvement, and as a vehicle for carrying cultural capital to the outside

Table 1 Opposite subjectivities

Other (working-class woman)	*Self*
(1)	
Old (or old-acting)	Young
Trapped	Free
Haggish	Glamorous
Ugly	Attractive
Fat	Slim/fit
Weak	Brave
Cog in machine	Autonomous
(2)	
Other (Alter-self)	
Trapped	Free
Pale	Tanned
Poor	Rich
Failure	Success
Depressed	Happy
Ordinary	Special
(3)	
Other (Eastern-European)	
Trafficked	Freely moved
Controlled	Independent
Desperate	Coping
Prostitute	Non-prostitute
Dangerous	Innocent
Oppressed	Liberated
Victim	Agent
Amateur	Professional

world. A way of accruing cultural capital is by improving the body, and 'distinction' (by way of comparison with unimproved bodies) reproduces improvement. For the lap dancer, the body is improved upon by regulation, not only by club rules (which dictate what she wears, how much bodily hair she can retain and so on) but also by herself as she adheres to the demands of an idealized femininity. The self-regulation is what holds her femininity together and what distinguishes it from other, degraded femininities. Despite her rigorous nightly workouts in the club, Rachel told me she still has to watch what she eats, because 'nobody likes a pot-belly'. As Skeggs notes, fatness signifies immobility, and 'the working-class body which is signalled through fat is the one that has given up the hope of ever improving' (1997: 83). What Rachel, a keen fan of Kentucky Fried Chicken (KFC) suppers after work, actually meant by watching what she eats is skipping meals when she feels she is gaining weight. Food in itself is not an important marker of taste, and is not regulated in the sense of it being a 'space of lifestyle' governed by approved or appropriate forms of food (Bourdieu, 1984). Bourdieu constructed a binary opposition between tastes of necessity and tastes of luxury (or freedom), the former being the tastes of the working-classes, the latter belonging to the bourgeoisie, for whom the tastes of necessity can be escaped by virtue of economic freedom. He also argued that when 'tastes of necessity' are maintained even when incomes are high enough to allow for freedom from them, this may constitute a form of 'cultural resistance', whereby tastes of necessity become liberating in themselves as ways of resisting refinement (Bourdieu, 1984). Rachel could easily afford to eat more luxuriously after work, yet she chooses not to. Bourdieu might have called this an act of 'congenital coarseness' (1984: 178); a process of spontaneous gratification and a show of solidarity to Rachel's cultural community. I think it indicates at least that investments are not being made towards emulating or become part of the 'middle-classes' by way of learning the signifying 'grammar' of food consumption (Barthes, 1979), and that this has wider relevance in terms of class (dis)association in this context. It is not about class aspiration *per se*, not least because all of the women without exception identified themselves to me, often with pride, as 'working-class'. Rather, as the data show, aspirations manifest as disassociation with very specific, unimproved (female) figures *within* the working classes.

So care of bodies is not undertaken by eating lightly, nutritiously, that is, fashionably, or by any process that is culturally representative. Rather, under-eating or meal skipping is done as a means to an end – a slim body shape. Bourdieu discusses the proportionality of interest in self-presentation and attention to the appearance of the (female) body to the material or symbolic value in it. Women, of different classes, he argues:

> [D]epend on the existence of a labour market in which physical appearance may be valorised in the performance of the job itself or in professional relations; and on the differential chances of access to this market in which beauty and deportment most strongly contribute to occupational value.
>
> (Bourdieu 1984: 202)

Lap dancing is certainly a form of work where physical appearance is particularly valorized in the 'performance' and has a direct relationship with the (symbolic and material) value and profitability of the body. Bourdieu discussed a logic whereby working-class women, whom he argues are less likely to work in jobs which 'most strictly demand conformity to the dominant norms of beauty' (e.g. professional/executive jobs), are also unlikely to be aware of the 'market value' of beauty and are therefore less inclined to 'invest time and effort, sacrifices and money in cultivating their bodies' (1984: 206). If it is a similar logic to this which informs the particularly pathologized representations of working-class femininity reproduced by Rachel, Karen and others, we can see that conforming to beauty norms allows the body to serve as an indicator of awareness of the value of their bodies. This awareness contradicts Bourdieu's assertion, and in subjective terms, enables the dancers to distance themselves from the lack of awareness of bodily value demonstrated by the working-class women back home. The care and attention bestowed upon bodies serves a professional purpose, because 'better' and better-kept bodies earn more lap dances and tips. However, it is also the *recognition* of this profitability that is important to subjectivity, because it reinforces hierarchies in (corporeal, cultural and economic) capital. The well-maintained body can thus become a vessel for distinction and respectability, social and symbolic profit as well as gaining control and making money.

spend, spend, spend

What the women do with the money they make with their bodies is also important to understanding their strategies for distinction. I asked the women about their consumption habits and a common theme to emerge was their reckless spending:

> I earn really decent money. I can have everything I want and I can eat out in a different restaurant every day. Some nights I get about €400. That's more than I could earn in a week, no, probably two weeks at home. Well, if I was doing just a crappy job anyway. People I went to school with think eating out is getting some fish and chips on a Friday night. Mind you, I should slow down 'cause it doesn't matter how much I earn, I always manage to spend every last penny.
>
> (Gemma)

Gemma's answer again sets up the contrast between herself and her past and the people she left there whom she views as having very limited aspirations and resources with regard to eating out. Dining in restaurants is beyond the limits of 'tastes of necessity', and to Gemma indulging in 'tastes of luxury' is connected to her new freedoms, especially in comparison to the Other, whose apparent lack of 'cultural capital' she mocks. Again though, not unlike Rachel with her inclination towards KFC, Gemma is either not fluent in, or chooses to reject the grammars of significance in food consumption:

> Me and my boyfriend go out in the daytime. Usually we have a full-English at about midday when we get up, then we'll probably spend some time together on the beach or in the flat. Sometimes I'll go to the club to practice on the pole. Then we go to the *Steakhouse* or for a Chinese buffet at about six o'clock. I eat so much 'cause of the dancing – you need it. My boyfriend's a bodybuilder so he eats a lot.

Natalie's money is also spent quickly and in ways that are instantly gratifying:

> I don't manage to save anything... it's not that easy to save out here 'cause there's so much you can spend your money on. Usually we leave work with a wad of notes but by the time you've bought some cocktails and some drugs, what you've got left doesn't go very far. You always need money for dresses and shoes as well. Anyway, it's nice to blow it when you've earned it.

Although dresses, shoes, cosmetics, perfume and other beauty products are necessary purchases for the dancers, there is an emphasis here on the act of spending for spending's sake. Katie echoed Natalie's attitude towards money:

> At first I couldn't believe the money I was making. I was going home with about €150 a night, and I had all these ideas about saving up and buying a car or getting a really posh apartment by the beach. But, I don't know... you just get used to earning a certain amount and it doesn't seem to go very far. Things are a bit quiet at the moment but I'm still earning easily enough to save up for something nice. It just doesn't seem to happen; I just seem to spend it as soon as I've got it. You get used to a certain lifestyle.

Eating in restaurants and drinking cocktails signal a surplus of wealth and this underscores the distinction between the women and their various Others. The idea that money can be wasted on 'conspicuous consumption' (Veblen, 1994/1899) is important to processes of dissimulation because it exemplifies another key difference between past and present. But wastefulness in this context may have another dimension.

This paper has so far been discussing how working as a lap dancer in Tenerife can facilitate the search for respectability, honour and extraordinariness. It has considered how these women produce themselves in comparison to other working-class British women (their former selves included) through a narrative of self-improvement, through the body and through lifestyle. Yet, unlike the group of women studied by Skeggs, who sought respectability in thoroughly 'respectable' institutions, that is, education, marriage, social care employment and (hetero)-sexual propriety, this group inhabits a rather different social and moral landscape.

Lap dancing, despite becoming an increasingly 'sanitized' and socially accepted form of striptease, still remains a relatively deviant career choice. Indeed the clubs visited during this study were a world apart from the luxuriously decorated and fiercely regulated *Spearmint Rhino* and other show-biz haunts featured in men's lifestyle magazine, which legitimize and glamorize demand as well as supply. Clubs X and Z were both raided and temporarily closed by the authorities on a number of health and safety breaches, and Club Z changed its policy from 'no extras' to 'extras supplied' when trade was slow.

Lap dancing's proximity to, and periodic crossover with, prostitution places it within some definitional, moral and legal contention. For example, is the performance of a lap dance to be included in the list of 'sexual services' exchanged for 'cash and/or other material benefit' typically understood as characterizing prostitution (O'Connell Davidson, 1998: 9)? How do lap dancing and other forms of 'exotic' dance further problematize the limitations of such definitions as identified by Phoenix (1995)? Can we regard lap dancing as belonging to 'the institution which allows certain powers of command over one person's body to be exercised by another... in order that he may command the prostitute to make body orifices available to him, to smile... to whip, spank, massage or masturbate him...' (O'Connell Davidson, 1998: 9–10)?

In clubs that operate strict and consistent 'no touch' and 'no extras' rules, the level of power exercised over lap dancers by customers to do such things is limited. The dancer has a degree of choreographic freedom, but she begins each shift knowing what is expected of her in terms of how far to strip, how close to get, how explicit the moves should be and so on. These standards are set 'in house'. Therefore, if a client requests anything further, for example, if he tries to touch the dancer, or make her touch him or herself, she can simply refuse and so his powers of command are overruled by those of the dancer's employer. Only in clubs where extras are provided are these terms negotiated between dancer (and in such cases prostitute) and client and therefore subject to his powers of command. Lap dancing, although it takes the body as the immediate site of labour, does not normally involve the 'intimate, messy contact with the (frequently supine or naked) body, its orifices or products through touch or close proximity' discussed by Wolkowitz [in her analysis of the social relations of 'body work' (2002: 497)].

That said, as mentioned above, club policies on touching and extras can sometimes be swayed by market forces. As well as Club Z reneging on its promise to dancers that they would not be asked to perform extras, Club X also rethought its rules when business was slack. During the course of less than four weeks in summer 2004, this club went from strictly enforcing a 'two-foot' rule (a distance of at least two feet must be kept between dancer and client at all times) to requiring its dancers to do 'enticement performances' including 'lesbian' shows (intimate dancing with touching and kissing) and a routine which involved a dancer covering her breasts and stomach in 'squirty cream' and allowing clients to lick it off. Dancers are not paid extra for these routines; they are simply intended to generate demand for lap dances.

Lap dancing, therefore, is not a homogeneous or clearly definable work form. It belongs to the sex industry and sometimes meets head on with prostitution. Yet, this chapter is arguing that lap dancing can also be the site for the accruement of capital and respectability. Although it is clearly a different kind of respectability being pursued by the dancers to that sought by the women studied by Skeggs, markedly so in relation to marriage, deferred gratification and occupation, respectability is still a significant feature of their narratives of self. So how might they begin to reconcile the contradiction between performing morally and sexually 'deviant' work with becoming respectable?

dirty money

First, in connection with wastefulness, the women become caught in a cycle of earning and spending, without ever saving money or investing in permanent material goods. For example, none of them owned an apartment or a car, or even their own furniture. In the main, everything earned in a night's work would be spent before work commenced the following evening, mainly on food, cocktails and drugs (usually cocaine). Several dancers said they often had difficulty paying their rent because they had failed to put it by. Natalie's propensity for 'blowing it once you've earned it' was more or less representative of the group.

It is conceivable to read these spending habits not only as conspicuous consumption for the benefit of status reproduction but also as a way of disposing of 'dirty money'. Hoigard and Finstead (1992: 49) identified a 'divided economy' among prostitute women in Oslo, wherein money earned legally was carefully budgeted and money earned from prostitution which 'burns a hole in your pocket and has to be used quickly' was spent on going out. Alcohol and drug use certainly serve other social purposes and can become habitual, but there is something in this type of compulsive spending that is reminiscent of Patricia Adler's (1993) ethnography of upper-level drug dealing in the USA, in which she writes about dealers feeling unable to spend their high earnings quickly enough. The dealers talked about wasting and burning money, and laughing when hundred dollar bills were found in laundered shirt pockets, and I want to argue that perhaps it is not simply its abundance that devalues money earned on 'deviant' markets. 'Marking' this money as dirty (see Zelizer, 1997) and shedding it as quickly as possible might also help to unpick the social meanings of how the money was made.

all good, clean fun

A second way to redress the contradiction between working in the sex industry and becoming respectable has to do with the discursive emphasis on drawing, redrawing and maintaining boundaries between different types of sex work. Again in the spirit of disassociation, integral to being a respectable lap dancer is *not* being a disreputable prostitute. The following three interview excerpts illustrate the importance of this distinction to the dancers themselves:

> I used to think lap dancing was pretty cheap, you know, slaggy... But now I know what it's about and it's not cheap. It's not like being a prostitute; I could never be a prostitute. ...No, we leave that to the Eastern Europeans. (Samantha)

> I know at [Club name] they've started making the girls do extras 'cause it's changed hands. [Colleague's name] used to work there but she came here when

> they told her she had to do extras... like giving blow-jobs and hand-jobs. I wouldn't do extras, never. I'm not a prostitute.
>
> (Natalie)

> Sometimes you have a bad night, like sometimes you get called a slag or a dirty whore by some dick-head in the crowd. You just ignore them. Mostly people are more respectful. ...I know I'm not a whore. I've never had sex for money and never would. It's not really the English girls who do that out here... mostly it's the Russian girls or the Eastern Europeans. We've had them here, and they'll do anything, they're so desperate...they're too desperate. It gives the rest of us a bad name.
>
> (Mel)

The lap dancers construct what they do for a living as a skilled, legitimate profession, entirely unrelated to prostitution and the sex industry. They experience their work as distinct from selling actual sex, which they consider cheap, desperate and non-respectable. Thus, they disconnect themselves from the dirt associated with prostitution and the universally constructed figure of the prostitute as a moral and sexual outsider (O'Connell Davidson, 1998: 129). National/racial distinctions are drawn within what constitutes prostitution (and therefore respectability), with 'Eastern European' and Russian women falling on the wrong side of the divide and British women remaining strictly on the lap dancing-only/respectable side. Rather than forming a politicized defence of the sex industry as a legitimate labour institution in which they participate, the women hold that lap dancing is not a part of the sex industry and that lap dancing/striptease have nothing to do with selling sexual services. The dancers themselves partake in the universal discourse which marks out prostitutes from their 'good sisters' in the Madonna/whore dichotomy (O'Connell Davidson, 1998: 131). Prostitution, like poverty, is a source of pollution associated with Other women, against whom the lap dancers contrast themselves. The contrast is further underpinned by its connection to the mechanics of racist self/other distinction. The Other woman is embodied by the racialized (and gendered and classed) identity of 'Eastern European' figure, who is at once feared, pitied and resented – feared for her abjection and ruthless drug-dealer boyfriend, pitied for her victimhood (having being trafficked against her will and beaten by her pimp), and resented for her desperation and willingness to sell sex and 'do extras', thus degrading the work of lap dancers. My wider research in Tenerife revealed that 'Eastern Europeans' have become demonized (and 'alienized' in Anglocentric disassociations with 'immigrants' and indeed with the act of migration itself – see Bott, 2004) by the British migrant community, and are blamed for a range of 'social problems', including human and arms trafficking, racketeering, drug dealing, theft and murder. Among lap dancers and other British migrants, they get the blame for prostitution too.

The Eastern European sex worker represents and embodies the polar opposite to the British lap dancer on many levels (see Table 1(3)). The key site for comparison and distinction is the way sex and sexuality are performed. Eastern European women sell sex and sexual services because they are in such 'desperate' situations that they have no choice. They are accused of degrading and de-professionalizing lap dancing. They represent a racialized sexuality which is coded by the British lap dancers as being unrespectable and which they quote in oppositional reference to their own sexualities. They transfer the dirt and immorality historically associated with their own classed, gendered and ethnicized social identities onto a new figure of sexual victimhood/transgression. By attaching her to, and blaming her for, whoring, they can detach themselves from such and claim some moral high ground.

conclusions

The young, working-class British women of this study share common experiences through being in Tenerife and working in the lap dance industry. They imagine, discuss and experience this new form of migrant labour in (largely) positive social–psychological terms, and engage self-consciously with it as physical, economic and social mobility. They discuss it as a means of escape and self-realization and encounter it as a way of disassociating themselves from that which degrades, disappoints and disgusts them.

Moving to Tenerife and working in lap dancing can be discussed as a process in which young British women negotiate themselves subjectively and seek respectability. They do this by engaging in what Skeggs (1997) called 'disidentification' and dissimulation with pathologizing versions of white working-class femininity. They disassociate with (but never forget) their 'alter' self – the woman they envisage they would have become had they not migrated, as well as with other working-class women living ordinary lives in Britain. They become valorized by *not* being these women. They work in very close proximity to prostitution, yet they are not prostitutes. The dancers stand out from the dirt and degradation of working-class femininity, not by following institutionally or traditionally respectable trajectories (marriage, occupation, family), but by claiming respectability in morally ambiguous ones.

While these analyses have limitations in terms of predicting the longevity of the embodied capital cultivated by the women which, because of the nature of the work has a potentially limited shelf-life, they can, however, contribute to understandings about negotiations of identity, especially in terms of how games of distinction are played out by those who cannot demand a great deal of recognition in the 'legitimate' status order. Different self-defined status groups distinguish themselves from, and disavow the attributes of, each other. Oerton and Phoenix's (2001: 389) analysis of the distinctions drawn between two often elided forms of body-work – sex work and therapeutic massage – argues that (because of the elisions) 'imputations of disreputability and immorality [are] made manifest'. It is the 'symbolic violence' of distancing oneself from others (Bourdieu, 1984) which unravels these imputations, in a kind of cultural pass-the-parcel, where groups repudiate the respectability of those who have taken a different trajectory to their own within the narrow range of possibilities facing young working-class women with similar holdings of social and cultural capital.

Lap dancing is a form of work in which a glamorized deviation from ordinariness and dirt can be discursively experienced and through which new sexualities and subjectivities may be imagined by dancers in their *own stories* of themselves and comparisons with others. In the migrant context in Tenerife at least, it is a location where certain pathologized femininities can become especially separate and other. It can be discussed as an institution wherein working-class women have some power to attach their own value to symbolic capital and claim legitimacy, or what Skeggs has termed a 'right to be' (1997: 11).

author biography

Esther Bott is Lecturer in Sociology at the University of Nottingham, where she previously worked as Research Associate on an ESRC funded project mapping and exploring and demand markets in the sex and domestic work industries of Britain and Spain. Her research and teaching interests include British migration to Spain, migration/tourism, social divisions and work, the body and self.

references

Adler, P.A. (1993) *Wheeling and Dealing: An Ethnography of an Upper-Level Drug Dealing and Smuggling Community* 2nd edn., New York: Columbia University Press.

Barthes, R. (1979) 'Towards a psychosociology of contemporary food consumption' in R. Forster and O. Ranum, editors, *Food and Drink in History*, Baltimore: John Hopkins Press, pp 166–173.

Bernard, C, De Gabrielle, C., Cartier, L., Monk-Turner, E., Phill, C., Sherwood, J. and Tyree, T. *et al.* (2003) 'Exotic dancers: gender differences in societal reaction, subcultural ties, and conventional support' *Journal of Criminal Justice and Popular Culture*, Vol. 10, No. 1: 1–11.

Bott, E. (2004) 'Working a working-class utopia: marking young Britons in Tenerife on the new map of European migration' *Journal for Contemporary European Studies*, Vol. 12, No. 1: 57–70.

Bourdieu, P. (1984) *Distinction: A Social Critique of the Judgement of Taste*, London: Routledge.

Castles, S. and Miller, J. (2003) *The Age of Migration: International Population Movements in the Modern World*, New York and Basingstoke: Palgrave MacMillan.

Cohen, R. editor (1995) *The Cambridge Survey of World Migration*, Cambridge: Cambridge University Press.

Enck, G.E. and Preston, J.D. (1988) 'Counterfeit intimacy: a dramaturgical analysis of an erotic performance' *Deviant Behaviour: An Interdisciplinary Journal*, Vol. 9: 369–381.

Forsyth, C. and Deshotels, T. (1997) 'The occupational milieu of the nude dancer' *Deviant Behaviour: An Interdisciplinary Journal*, Vol. 18: 125–142.

Frank, K. (1998) 'The production of identity and the negotiation of intimacy in a gentleman's club' *Sexualities*, Vol. 1, No. 2: 175–201.

Frank, K. (2002) *G-Strings and Sympathy: Strip Club Regulars and Male Desire*, Durham and London: Duke University Press.

Hoigard, C. and Finstead, L. (1992) *Backstreets: Prostitution, Money and Love*, Cambridge: Polity Press.

King, R. (2002) 'Towards a New Map of European Migration' *International Journal of Population Geography*, Vol. 8: 89–106.

Lewis, J. (1998) 'Learning to strip' *The Canadian Journal of Human Sexuality*, Vol. 7: 51–66.

O'Connell Davidson, J. (1998) *Prostitution, Power and Freedom*, Cambridge: Polity Press.

Oerton, S. and Phoenix, J. (2001) 'Sex/bodywork: discourses and practices' *Sexualities*, Vol. 4, No. 4: 387–412.

Phoenix, J. (1995) 'Prostitution: problematizing the definition' in M. Maynard and J. Purvis (1995) editors, *(Hetero)Sexual Politics*, London: Taylor & Francis.

Skeggs, B. (1997) *Formations of Class and Gender: Becoming Respectable*, London: Sage.

Steedman, C. (1986) *Landscape For a Good Woman: A Story of Two Lives*, London: Virago.

Veblen, T. (1994/1899) *The Theory of the Leisure Class*, New York: Dover.

Walkerdine, V. (1990) *Schoolgirl Fictions*, London: Verso.

Walkerdine, V. and Lucey, H. (1989) *Democracy in the Kitchen: Regulating Mothers and Socialising Daughters*, London: Virago.

Wolkowitz, C. (2002) 'The social relations of body work' *Work Employment and Society*, Vol. 16, No. 3: 497–510.

Wood, E.A. (2000) 'Working the fantasy factory' *Journal of Contemporary Ethnography*, Vol. 29: 5–31.

Zelizer, V.A. (1997) *The Social Meanings of Money: Pin Money, Paychecks, Poor Relief and Other Currencies*: Princeton University Press.

doi:10.1057/palgrave.fr.9400279

83

female sex tourism: a contradiction in terms?

Jacqueline Sanchez Taylor

abstract

This paper argues that the 'double-standard' applied to male and female tourists' sexual behaviour reflects and reproduces weaknesses in existing theoretical and commonsense understandings of gendered power, sexual exploitation, prostitution and sex tourism. It looks at how essentialist constructions of gender and heterosexuality blur understandings of sexual exploitation and victimhood and argues that racialized power should also be considered to explore the boundaries between commercial and non-commercial sex. This paper is based on ethnographic research on sexual–economic exchanges between tourist women and local men and boys in the informal tourist economy in Jamaica and the Dominican Republic.

keywords

sex tourism; sexual exploitation; heterosexuality; racism; gender

(42–59) 0141-7789/06 $30 www.feminist-review.com

introduction

Tourism is a world growth industry and one of the fastest, if not only, growing industries in the Caribbean. Jamaica and the Dominican Republic are two of the biggest tourism centres in the region and are heavily reliant on tourism as a source of foreign exchange. In 2004, the Dominican Republic welcomed over 3 million visitors and Jamaica, one of the poorest countries in the region, welcomed 1.3 million, continuing a pattern of steady growth since the 1990s. Both the Dominican and Jamaican governments spend on average up to 9% of total government spending on tourism and travel infrastructures (WTTC, 2005). Tourism has become central to economic development programmes designed to reverse crippling economic problems and poverty, repay international debts, and improve rates of unemployment (on average 15% in Jamaica and the Dominican Republic). However, the International Monetary Fund agreements and World Bank structural adjustment loans, sector adjustment loans and programmes loans that the Jamaican and Dominican governments entered into have also had a devastating impact on the poor. The policy packages tied to these loans have undermined traditional forms of subsistence economies and redirected subsidies away from social spending and basic commodities towards debt servicing, and adjustment processes have also involved massive currency depreciation and a concomitant drop in the price of labour. Unemployment rates have risen, and the poor in Jamaica and the Dominican Republic have increasingly been forced to scavenge a living in the informal economy (Anderson and Witter, 1994; Kempadoo, 1999).

Prostitution and other forms of tourist–local sexual–economic exchange are among the wide range of activities that take place in the informal tourism economy and even workers employed in the formal tourism sector sometimes supplement low wages by entering into sexual–economic exchanges with tourists. Increasingly, it is not just local women who are entering into the informal tourist sex industry but also men predominantly providing sexual services for tourist women. One consequence of introducing affluent tourists to poor local communities has thus been the development of tourist-related prostitution, a phenomenon that leads both male and female sex tourists to describe such places as 'sexual paradise' and 'Fantasy Island', places where sexual dreams come true (O'Connell Davidson and Sanchez Taylor, 2005).

And yet a 'double-standard' is often applied to male and female tourists' sexual behaviour in academic commentaries on the phenomenon, with male tourists being described as 'sex tourists' but female tourists being described as engaging in 'romance tourism'. Drawing on ethnographic research on sexual–economic exchanges between tourist women and local men and boys in the informal tourist economy in Jamaica and the Dominican Republic, this article argues that female tourists' sexual–economic relationships with local men are predicated upon the same global economic and social inequalities that underpin the phenomenon of male sex tourism. The fact that parallels between male and female sex tourism are widely overlooked reflects and reproduces weaknesses in existing theoretical and commonsense understandings of gendered power, sexual exploitation, prostitution and sex tourism.

reading female tourists' sexual—economic relationships with local men

The stereotypical image of the 'sex tourist' is that of the Western man who travels to Thailand or the Philippines in order to engage in brief, highly commodified exchanges of sex for cash or kind with prostitute women or children (Enloe, 1989; Jeffreys, 1997). Because 'sex tourism' is taken to imply prostitute-use, and because prostitute-use is assumed to be a male practice, it is difficult to conceive of 'female sex tourism', which becomes almost a contradiction in terms. Although researchers often acknowledge that sexual relationships between local men and tourist women are based on an exchange of money or goods and gifts, they are generally unwilling to apply the terms 'prostitution' or 'sex tourism' to this phenomenon (for example see Momsen, 1994; Pruitt and LaFont, 1995; Yuval-Davis, 1997). This is not simply because the actors themselves tend to describe their relationships through reference to the concept of romance rather than prostitution, but also because essentialist models of gender and sexuality, and dominant understandings of the term 'sexual exploitation', preclude the possibility that a woman can sexually exploit a man (Sanchez Taylor, 2001).

Research shows that the term 'male sex tourism' embraces a wide range of different activities and relationships spanning brief, explicit cash-for-sex exchanges through to longer term relationships, within which, 'romance' is traded for economic support and other benefits (Seabrook, 1996; O'Connell Davidson, 1998; Brennan, 2004). At this end of the spectrum, male tourists do not always perceive themselves as 'sex tourists' (Kleiber and Wilke, 1995; Gunther, 1998). Even when male tourists do not read their own sexual behaviour as participating in prostitution, academic commentators are usually happy to refer to them as 'sex tourists'. This is not the case with regard to women tourists who behave in identical ways. Instead of being understood as problematic, such relationships are represented as involving a mutual exchange where the local men are constructed as equal if not more powerful in relation to tourist women (Meisch, 1995; Pruitt and LaFont, 1995; Dahles and Bra, 1999; Phillips 1999). Indeed, Jeffreys contends that it is vital to maintain the distinction between men's sex tourism and women's romance tourism as it is in her view politically problematic to degender prostitution by arguing 'women do it too' (Jeffreys, 2003: 225).

North American and European feminist debate on prostitution largely centres on the question of whether the female prostitute is a victim of male sexual violence (e.g. Barry, 1995; Jeffreys, 1997), or whether there can be 'free choice' prostitution within which the prostitute woman is an autonomous agent choosing to sell her sexual labour (see Chapkis, 1997; Doezema, 1998; Kesler, 2002). Male prostitution does not usually feature in these debates, partly because it is assumed that male sex workers provide sexual services mainly to other men and that therefore questions of gendered power do not pertain. Altman argues that this results in a 'belief that homosexual sex work is inherently less exploitative than heterosexual sex work' (1999: xiv) because a level of equality is assumed to be present between male client and male seller. As Kesler further notes 'male prostitutes, or for that matter promiscuous men, are not seen as 'selling their humanity' when they engage in sex with many partners, for money or for free' (2002: 226), and so debates on prostitution and exploitation have largely excluded them from the analysis. But male prostitution has also been overlooked because victimization is a highly gendered concept (see Hakken, 1999; Lamb, 1999), making it difficult to discuss ways in which men might be sexually exploited by women.

Thus, although North American and European feminist debate on prostitution includes a number of different perspectives on gender and sexuality, analyses of the sex trade as a site of *exploitation* tend to be informed by a radical feminist perspective, and this generates two key problems in relation to the phenomenon of female sex tourism. First, in radical feminist analyses, the terms 'sexual exploitation' and 'sexual victimization' tend to be conflated, and it is therefore hard to speak of any individual as 'sexually exploited' without also implying that they are downtrodden, pathetic and suffering. Commentators such as Agustín (2003) have been very critical of this construction of victimization. The term 'victim' not only feminizes the person who exchanges sex for some economic benefit, but also implies a total lack of agency on that person's part, as though she is simply the passive puppet of her circumstances. The language of 'victims' and 'victimizers' makes it seem that the power relations within sexual economic exchanges are identical to those which exist in cases of rape or incest (see Barry, 1995), and so also constructs the prostitute's client as an active abuser and therefore necessarily a man.

Second, because radical feminist analyses of sex tourism start from the assumption that it represents a form of prostitute use and can therefore be primarily explained in terms of patriarchal power relations, very little attention is paid to the racial and global power relations that underpin the phenomenon. Rather than thinking about the complex intersections of race, gender, class and nationality, there is a tendency to simply tag notions of racial or North–South inequality on to the explanatory framework (Enloe, 1989; Barry, 1995; Jeffreys, 2003).

More recent research on sex tourism examines the impact of globalization and international economics on global inequalities and identities, and explores the dynamics of power relations between tourist men and local women more fully (de Alburquerque, 1999, 2000; Wonders and Michalowski, 2001; Brennan, 2004; Cabezas, 2004). However, the power relations between local men and tourist women have not been explored in such detail. There is still a tendency to focus on gigolos as agents who either economically, emotionally or sexually exploit tourist women rather than being exploited by them. This suggests that we need a more complex model of gender power and sexuality than that currently offered by North American and European feminist commentators on prostitution, such that it would be possible to speak of local men who engage in the informal sex industry (and others who are involved in the immense range of types of sexual economic exchange in the contemporary world) as being exploited without necessarily also having to think in terms of passive victims and malicious victimizers, and to explore the significance of race for their experience of exploitation.

Recent Caribbean feminist scholarship on gender, race and sexuality promises the beginnings of such a model (for example Mohammed, 1998, 2003; Barriteau, 2003; Lewis, 2003; Sheller, 2003). Critical of North American and European feminist theory and debate, which universalizes women's experience, such theorists are concerned with the specificity of Caribbean women's experience and the ways in which the dynamics of race and gender have historically been shaped by global relations of power (Mohammed, 1998).

Kamala Kempadoo's (1998, 2004a, b) valuable work tracing the intersections of sexuality and race in the Caribbean on sex work stresses the ways that sex in the Caribbean carries an important economic value and has been exchanged and used as a form of resistance against poverty, marginalization and exclusion. Kempadoo argues that very often Western feminist accounts of sexuality ignore 'a continuum of sexual relations from monogamy to multiple sexual partners' and the fact that sex may be viewed as an asset that a woman can trade 'in favour of specific western ideologies and moralities regarding sexual relations' (Kempadoo, 1998: 12, see also de Zalduondon and Bernard, 1995). Although Caribbean feminist scholars have focused primarily on the experience of women in their analyses of sexuality, race and gender, they have also noted that Caribbean men sometimes use the exchange value that attaches to sex (especially the sexual value attached to the black male body) in the same way. In this way, they stress that Caribbean men and women need to be recognized as having been both 'the objects as well as the subjects in writing the sexual script in the Caribbean' (Kempadoo, 2004b: 182).

Important as it is to explore ways in which Caribbean women and men actively author aspects of the sexual script in the Caribbean, it is also necessary to recognize that global inequalities often give a great deal of 'editorial control' to

North American and European tourists who are party to sexual–economic exchanges with locals in the Caribbean. And if questions about tourist women's motives for entering into sexual relationships with local men and questions about what is 'sold' by the local men are examined in more detail, then sexual economic exchanges between female sex tourists and local men begin to look much more problematic than either Western or Caribbean feminists' accounts suggest.

women, sex and power

Radical feminist theory suggests that sexuality is a key site of male power. In 'the sex act', women submit to men, and men affirm their masculinity and patriarchal power by penetrating the female body (Dworkin, 1979; Pateman, 1988). This model of gender power as domination constructs relations between men and women as a master–slave relation and therefore treats both men and women as undifferentiated groups and obscures the significance of class, race, and age for an individual's social power and life chances (see Brace and O'Connell Davidson, 2000). This model also conceals the fact that women too can feel empowered by certain aspects of heterosexuality and are sexual agents.

Sexual mores at the turn of the millennium are, as Sue Scott and Stevi Jackson (2000: 176) observe, 'confused, contradictory and contested'. Nevertheless, most societies remain deeply committed to the idea of gender difference and idealized forms of femininity and masculinity continue to be promoted and valued. Thus we find that although women's economic, political and social status have undergone many changes in Western countries during the twentieth century, most Western women still wish to be seen as 'feminine' and would find it insulting to be described as 'masculine'. As Bordo (1993) notes, women collude to maintain femininity because social honour is bound up in accepting the confines of their gender.

Women's gender identity, and so their gender honour, is also strongly linked to sexuality. Most discussions of women's gender honour have focused on its traditional links to sexual purity and passivity (a woman has traditionally de-gendered, and so dishonoured, herself by taking a 'promiscuous', active or instrumental approach to sexual life). However, women are nonetheless required to have a heterosexual sexual identity in order to attain gender honour, so that investing in social ideals of 'femininity' means investing in social ideals of female heterosexuality. Since social honour is awarded to those who uphold the idealized attributes of their perceived gender roles, heterosexual women who are considered undesirable to men, or who are unattached, as well as lesbians, are often at risk of losing honour within their community in just the same way that men who do not conform to masculine norms are often dishonoured.

In contemporary Western societies, social ideals of 'femininity' still seem 'to reside in the female body and the female heterosexual response' (Sonnet and Whelehan 1995: 83). The women who are celebrated in advertisements, magazines, film and fiction are mostly, young, sexy and able to command the gaze of men. Large numbers of Western women pay surgeons and cosmetic companies for treatments and products that will give them a more youthful and feminine look. Heterosexual sexual imagery is also everywhere in popular culture and used both to sell and to entertain. Although story lines of television shows like 'Sex in the City', 'Desperate Housewives' and novels like *The Diary of Bridget Jones* (Fielding, 1996) seem to reflect a concern with changing gender roles, they are also entirely preoccupied by the neurotic search for heterosexual love and relationships. Meanwhile, gossip magazines and newspapers continue to provide a constant parade of stories about the marriages and real-life romantic relationships of the stars, and their experiences of childbirth and motherhood, so that all can collude in and celebrate their heterosexuality.

Women are simultaneously encouraged to be economically independent (a number of recent hit songs explicitly call on women to take pride in making their own money, paying their own bills, etc.), and yet to continue to value traditional ideals of gender difference in order to become good wives and mothers. In other words, they are encouraged to buy into a contradiction. There is little concern to challenge the patriarchal power structures and the traditional emphasis on gender difference is re-worked and re-inscribed in contemporary Western popular culture (for example, in best selling books like *Women are from Venus, Men are from Mars* (Gray, 1998), and films like *What Women Want* (director: Meyers, 2000)).

Against this backdrop, heterosexual sex remains a medium for social affirmation for women as well as men. Unless a woman is publicly known to be being 'sexed' by a man or men, there is a question mark over her 'femininity'. Because ideals of heterosexual sex are still powerfully linked to the social taboo against the sameness of men and women, 'real' sex is still taken to imply penetrative sex involving male 'activity' and female 'receptivity' and the act of penetration is still taken to simultaneously engender both parties (Kitzinger and Wilkinson, 1993; Jackson, 1997). At the same time, however, discourses about gender difference turn sexual relationships between the supposedly 'opposite' sexes into difficult and painful emotional terrain. Because individuals are expected to become 'men' and 'women' through sexual intimacy with someone who is imagined as their opposite, someone different, heterosexual relationships are often constructed as inherently conflictual, an arena within which 'the battle of the sexes' is fought out. And yet 'sex difference', gender roles and the tensions it creates, is eroticized and romanticized to fuel sexual desire and romantic longing. The ideal man prizes and embodies a masculine identity that is constructed as above all not womanly (e.g. tough, violent, independent), and yet somehow also manages to value woman and female virtues such as tenderness, gentleness and devotion above all else. He is, in other words, also a contradiction, simultaneously different and the same, and it is not surprising that in real life, few women manage to find him.

These inconsistent discourses of gender, sexuality, romance, love and heterosexuality interlock with ideas about racial difference, and are of great significance for the tourist women who enter into sexual–economic relationships with local men in the Caribbean (O'Connell Davidson and Sanchez Taylor, 2005). Female sex tourists are looking for 'real men', rather than a 'New Man', and black men are perceived as being hyper-masculine. Homophobic sex laws and attitudes in Jamaica and homophobia in the Dominican Republic bolster this fantasy of black men as 'real' men, while at the same time racism also constructs Caribbean men as being closer to women, in the sense of being supposedly closer to nature, more intuitive, irrational and emotional than white men. Through the lens of racism, then, Caribbean men epitomize the romantic ideal – they are more like women, even as their 'animalistic' attributes make them more like men than white men.

In many Third-World tourist resorts, female sex tourists find themselves presented with opportunities to 'attract' these 'real men', and so to personify the ideals of femininity and heterosexuality that they are taught to aspire to. Equally, if not more importantly, they can find a 'real man' to sex them (thereby publicly affirming their 'femininity') without losing control and becoming a real woman in the sense of being a social and political subordinate or risking being rejected and humiliated.

An American woman, who was a regular visitor to Jamaica, summed up a popular recurring view of many women interviewed and was explicit about the control she enjoyed:

> Women who come [to Jamaica] like control. They don't need a man in their lives except for sex, don't need them for money - just sex – because they can't do that for themselves. Better to have someone uneducated that acts like a lackey and in your control, so you can say when it starts and when it stops.

Female sex tourists I interviewed generally spoke of feeling powerful in relation to local men, and some women also described feeling empowered in relation to white men, for in the Caribbean, where they could command the sexual attentions of black men, white men no longer had the power to control or reject them sexually. The kind of control exercised in their relationships with local men is actually very similar to that exercised by male sex tourists in sexual economic relationships with local women (O'Connell Davidson and Sanchez Taylor, 2005). On the whole tourist women do not enter into sexual relationships with middle-class men who would be their economic equals or superiors, but enter into sexual encounters with men working in the formal or informal sex industry. As affluent

Western tourists, they are able to use their economic power to limit the risk of being challenged or subjugated.

It is, for example, the female tourist who books the flights and determines the length of time they will spend with their boyfriend, as well as making day to day decisions when they are together, such as when and where they eat. Their boyfriends, however, are unable to pay for flights to visit them in Europe or North America, cannot get a visa to enter their country unless they vouch for them, and have little control over the details of their interactions. As a Jamaican hotel worker who provided sex to female tourists to supplement his income noted:

> Well, she tell me straight up front that she have three kids and she don't want to get involved. We could do this, we could do that. She don't want no personal relationship. One day you don't hear from me, things happen, you must take it just like that because it's not a long term relationship, ya know'

Although most of the women I interviewed and surveyed labelled their sexual encounters as 'romance', others were not ashamed to call their relationships 'purely physical' with 'no commitment', 'no chase'. For example, a 42-year-old English woman was unambiguous about why she travelled at least three times a year to the Dominican Republic for sex with local men in the popular tourist resort of Boca Chica:

> I'm not naïve, I've been around the block. I come for sex, of course the sun, but mostly the sex. I'm not coming to live and set up house with a guy I just want some fun and good sex.

And yet none of my respondents used the term prostitution to describe their encounters and none considered their sexual behaviour a form of sexual exploitation even when they acknowledged that it was 'all about money', and this was because they employed gendered constructions of sexuality to read their own sexual encounters. Many female sex tourists criticized male sex tourists for 'taking advantage' of the poor local women, yet did not see their sexual encounters as also taking advantage of global economic structures that empowered them to enter into very particular kinds of sexual relationships with local men.

problematizing male agency

Researchers who have examined sexual–economic relationships between tourist women and local men in developing countries have tended to concentrate on the men's agency and as such have described them as entrepreneurs, cultural brokers, or players who commodify aspects of their racialized or ethnic identity in order to make a living (Dahles and Bra; 1999; Phillips, 2002; Lewis, 2004). But agency does not necessarily equate to power in the sex trade, especially if the 'sex worker' or 'prostitute' does not subjectively interpret his or her sexual activities as a form of work. Entering into diffuse and long-term sexual relationships with tourist women can provide urgently needed economic support but often at great emotional costs for men who reject the sex worker label and have few boundaries limiting the level of intimacy during sexual acts they perform. Some men expressed a sense of rejection, disappointment, betrayal and dishonour. As one male interviewee in Jamaica said:

> Some ladies, they just come for two weeks, just for satisfaction. To satisfy their needs and their wants and they don't remember you. Many guys are treated like that. Like they use them or something. Lots of times I see this happen and feel this.

Both Jamaica and the Dominican Republic are societies where male honour is highly sexualized and policed. Men who felt dishonoured or used had no real framework to understand the motives of the tourist women who 'dumped' them because they did not want to acknowledge economic or social inequalities between themselves and the tourist women. This was the case even when the men were acutely aware of the way that racism and poverty restricted their economic opportunities, and how ideas of difference played an important role in tourist women's desire for sex with them.

Local men who took a more professional approach to their sexual encounters with tourist women, however, like professional female sex workers, made an instrumental distinction between sex and love (De Moya and Garcia, 1999; Herold *et al.*, 2001). As a Dominican man explained 'sex is not for pleasure but a business, I don't care if the woman is beautiful or not. I'd rather go out with someone ugly if she's got money'. This man worked on the beach for six years and made enough money to set up a little business. Although this small group of experienced male sex providers may take a very businesslike approach to their activities, they still have to perform actions that would be unacceptable to many North American or European sex workers, such as kiss their clients, hold hands, open doors, massage their backs, buy drinks, fetch and carry shopping and not ask for payment up front. One 21-year-old migrant from Haiti who had been working in Sosua, complained that he even had to 'snog' his tourist client despite a bad toothache and a swollen face, because if he did not, he would not be able to afford the antibiotics to cure it. This illustrates how the heterosexual codes of sexuality mean that the fictions of romance still govern sexual transactions taking place between tourist women and local men who enter into sexual economic exchanges with them. This makes it more difficult for the local men who provide sexual services to tourist women to control the transactions that take place within the sexual exchanges and hinders attempts to turn such relationship into work.

Exploring the motivations and attitudes of female sex tourists in Jamaica and the Dominican Republic suggests that gender power within heterosexual unions is highly complicated and variable and cut across by class, race and age, such that in some circumstances, women can pursue a social ideal of heterosexuality without automatically placing themselves in a subordinate position. This also has implications for theoretical models of sexual exploitation. If women are not necessarily subordinated by the heterosexual sex act, then it becomes possible to recognize that they too can, in certain circumstances, sexually exploit men.

The Longman English Dictionary defines 'exploit' as 'to use or develop fully, especially for profit or advantage... to take unfair advantage of for financial or other gain'. Julia O'Connell Davidson (2001: 2) observes that this emphasis on unfair advantage 'points to the existence of some imbalance of social, political, economic and/or physical, psychological or emotional power between the exploiter and the exploited'. Applied to the concept of sexual exploitation, this would imply that one party to a sexual interaction took advantage of an imbalance of power to obtain a sexual advantage that would otherwise have been denied them. It does not necessarily imply that the exploiter used violence, nor that the exploited would necessarily subjectively feel victimized, violated or exploited.

Using the term 'exploitation' in this way, I would argue that female tourists who travel to Jamaica or the Dominican Republic for 'romance' or sex do exploit Caribbean men in the sense that they wittingly or unwittingly take advantage of unequal global and local power structures in order to both pursue their own sexual pleasure, and to affirm themselves as raced, sexual and engendered beings and so to improve or consolidate their position on status hierarchies of gender, race and sexuality. And I will argue below that the power relations that underpin and are reproduced by female sex tourism appear even more troubling if what is sold by local men is interrogated in more detail.

what's race got to do with it?

Black bodies are increasingly used as props in advertising and media and younger white Europeans and North Americans, are constantly exposed to exoticizing racisms through the Western film, music and fashion industries (Hooks, 1992). These industries retain the old-school racist emphasis on blackness as physicality, but repackage and commoditize this 'animalism', so that black men and women become the ultimate icons of sporting prowess, 'untamed' rebelliousness, 'raw' musical talent, exotic beauty, sexual power and so on. As a consequence, many young (and some not so young) white Westerners view blackness as a marker of something both 'cool' and 'hot'. In certain contexts the black body is not feared but fetishized. Thus, some female sex tourists want black boyfriends in order to live out certain fantasies, whether they be 'educating and helping the noble savage', being the focus of 'cool' black men's adoring gaze or wanting to have 'sunshine babies'. Archer-Straw (2000) argues that although black people may enter into sexual and intellectual relationships with white people as equals, it must also be remembered that these relationships are grounded in existing economic and racialized inequalities and are dependent upon how individuals relate to the structures that they are located within.

Many of the female sex tourists I interviewed constructed their sexual relations with black men as a sign of anti-racism, yet because they did not think very deeply about racism, they also continued to espouse ordinary, 'everyday' forms of racism even as they denounced overtly white supremacist politics. For example, Sarah was a 32-year-old woman from Devon, who had travelled all over the world to 'experience other cultures' and 'find herself'. She had visited Negril several times and had a number of sexual relationships with local men, and she saw herself as a great liberal on questions of race. She told me that she was 'colour-blind', as colour did not matter to her. But she complained bitterly about how Jamaicans were racist towards white people. She told me that some had been 'hostile and aggressive' to her and called her 'whitey'. She concluded from this that 'we are all racist', and could not see any distinction between personal prejudice and institutionalized patterns of racist oppression and exclusion.

Williams (1997) argues that 'the very notion of blindness about colour constitutes an ideological confusion at best, and denial at its very worst' (1997: 2), and female sex tourists are often very confused, naïve or ignorant about what constitutes racism. Many have no concept of what racism is (outside of their own experience of racism in Jamaica as white women) or any notion of how racism operates.

'Everyday' racism of the type that asserts there are natural differences between 'races' and 'cultures' is widespread and common in the tourist industry, which uses notions of 'difference' to market long haul holiday destinations. People who promote tourism and run tourist business in the West use such understandings to construct Others and to interpret their own relation to them. Most tourism companies would also label themselves anti-racist but continue to reproduce the ideas about 'difference' upon which racism rests. Consuming 'difference' is presented as an acceptable, even desirable, aspect of tourism. Thus, Jennifer Cox, a contributor to the Lonely Planet Guidebook series, could appear on a radio program to defend the phenomenon of 'female sex tourism':

> I think that all the social cues are different. For example, a white woman may not get involved with a black guy in this country so it's a matter of not having an opportunity but it's also the sense that travel, that anything is possible, and that's one of the reasons why people travel. You want to go out and feel open to experiences and you want more than life at home and one of those experiences might be to get involved with a sexual partner that is very different from the one that you normally chose at home. It isn't necessarily racism, it might actually be the opposite to racism, a sense of real openness.
>
> (Jennifer Cox, Lonely Planet, Woman's Hour, BBC Radio 4, 1999).

Although commentators like this imagine that black and white coming together in tourist settings in the Caribbean is 'the opposite of racism', it could also be argued that this can only happen because the structures and ideas which lead to black people's social, economic and political marginalization are left unchallenged while black bodies continue to be erotically valued.

Analyses of sexual exploitation and sex tourism need to pay far greater attention to the links between sexuality and racism. In his discussion of Fanon's (1986) seminal discussion of 'negrophobia' Dollimore (1991: 345) observes that he 'is surely right to stress the sexual component of racism, especially its destructive effects on (hetero)sexual relations across race'. Dollimore further indicates that Fanon was also correct to stress 'the white person's fear of and fascination with the imagined sexual potency of the Negro' (Dollimore, 1991: 345). And yet because Fanon was so heavily invested in the idea of gender difference, and failed to critique gender hierarchies in the same way he critiqued hierarchies of race, he identified masculinity as a source of resistance, and thereby perpetuated 'in terms of sexual and gender relations, the very oppression being resisted at other levels' (Dollimore, 1991: 347). This kind of sexist thinking very easily translates into a celebration of the black phallus (Hooks, 1994). The paradox is that in celebrating the power and strength of the black phallus, black men who are economically and socially marginalized play into rather than resist white racist stereotypes about the black male as hypersexual. Thus, local men in the Caribbean who enter into sexual economic exchanges by acting up to the stereotype of the 'black stud' may find they are rewarded with kudos amongst their peers as well as material benefits, but they also reinforce and perpetuate the racist stereotype. So, for example, local men's chat up lines include frequent reference to Jamaica or Dominican men's sexual prowess and are littered with claims about having large penises and being naturally great or insatiable lovers. Female tourists then repeat these ideas, stating that *all* Jamaican or *all* Dominican men or *all* black men 'just love women' or are 'obsessed with their dicks' or 'think of nothing but sex'. Since local men's economic and social marginalization is inextricably linked to racism (both contemporary and historical), there is a sense in which, in order to survive economically, they have to contribute to their own continued oppression. Researchers have linked the phenomenon of male sex tourism to historical colonial constructions of the exotic sexual object (Yuval-Davis, 1997), but it is equally important to trace connections between colonial racist constructions of black/Other male sexuality and the sexual practices of female tourists in poor and developing countries.

conclusion

Gunther (1998) argues that researchers should look carefully at the fuzzy line between commercial and non-commercial sex that is continually drawn and redrawn by tourists and local people who enter into sexual relationships. He warns us that 'by idealizing sex tourists as romanticists who travel in pursuit of emotional relationships, one may easily overlook the fact that these romanticists use and reproduce sex tourism as a social institution' (1998: 80). Thinking about the relationship between sex tourism, tourism and inequality may also help theorists to better understand questions about sexual exploitation.

When North Americans and Europeans take holidays in poor and developing countries, they benefit from a particular and highly unequal political and economic world order that shapes their relationships with local people. Because so many poor and uneducated local people are denied the privilege of entering into the formal tourist economy, they are forced to survive on the margins of the informal sector. The tourist industry sells long haul holidays as 'an experience', and as such, it relies on local people to provide a face and character to what would otherwise be a standardized hotel, beach and holiday for Western tourists. It therefore depends on the informal tourist sector, which operates along side the formal industry, to provide the 'local colour' that cannot be supplied by hotel employees and tour representatives.

Local people's informal economic activity is therefore very important to the tourism economy, for though not all tourists want to step outside of the staged tourist experience provided by large hotels and all-inclusive resorts, those who do rely on informal sector workers to supply them with a taste of the 'authentic' Caribbean. MacCannell (1976) notes that in tourism, there is a continuum from front stage to back stage reality, rather than a clear and sharp divide. Often, the back regions of tourist resorts that tourists are allowed to enter into are fake and like a stage behind the stage, one step along the continuum. This fake back region of tourism in the Caribbean is where much informal economic activity takes place. It provides earning opportunities to those who are excluded from the formal sector by lack of connections, education, or gender or race discriminatory policies. The back stage (which tourists believe holds the secrets and truth of real, authentic Caribbean life) is largely controlled by local people who are excluded from the formal tourism economy.

One of the key roles of men who enter into various forms of sexual economic exchanges in particular, is to provide access to this region, and tourists generally believe that they enter into more 'genuine' relationships with 'the locals', who are assumed to be off stage, than with hotel employees who are assumed to be on a public stage. In reality, however, local men who enter into sexual relationships with tourist women as much as workers in the formal tourism economy are engaged in an economic (and so an instrumental) relationship with the tourists they befriend, and not pursuing intimacy for personal reasons. Tourists may want to believe that their relations with workers in the formal and informal tourism economy are different, but in reality, the relations of power that exist on the front stage are also 'woven into the fabric' of the back regions (MacCannell, 1976: 91).

In the Caribbean, these relations of power are both economic and racialized. The female sex tourist enters the back region from a position of privilege, and to satisfy personal desires rather than to meet economic needs. She is free to exit that region at any point. For local men, the back region is part of the 'economy of makeshifts' (Brace, 2002) that they rely upon for economic survival, and they cannot freely move out of it. Furthermore, because tourists turn to the back region in pursuit of some 'authentic' and 'genuine' experience, those who work in the informal sector often have to trade in things that are not usually considered saleable, namely friendship, intimacy, and/or sex. Although men involved in tourist related sexual exchanges and their female clients are often equally concerned to ignore or downplay the material inequalities that divide them, and to construct their relationships as if they were based simply and solely on mutual desire and/or romance and/or friendship, their motives for constructing a fiction are different, and have different consequences. Black Caribbean men who provide sexual services as an economic activity are performing their 'Otherness' in ways that are desired by tourists. As discussed above, this means reproducing and perpetuating racist stereotypes about the Other and having to subjectively experience sexualized racisms as empowering. The tourists' economic survival does not depend on them performing any particular version of 'femininity', and tourist resorts in the Caribbean actually provide them with a stage on which to perform their femininity in any way they choose.

Nor should it be assumed that men participating in the informal sex tourist industry always subjectively experience their participation in the back region as affirming their own vision of masculinity. Because the back region rests on the creation of a range of fictions, fantasies and illusions, it can be an emotionally dangerous place for local men. There is a tension between providing a convincing performance, and coming to believe that the performance is real. Local men, like local women involved in the sex industry, can experience a sense of personal rejection when the fact that female sex tourists are merely using them as objects is made explicit.

My research suggests that to understand sexual economic exchanges between tourists and locals in tourist areas in the poor and developing world, it is necessary to move away from a gender essentialist understanding of sexuality and exploitation and focus instead on power in its multiple configurations. If we make this shift, female tourists' sexual encounters with local men in poor and developing countries do not look so very different from male tourists' sexual encounters with local women, and 'female sex tourism' no longer appears as a contradiction in terms.

acknowledgements

This research formed part of a broader Economic and Social Research Council (ESRC) funded ethnographic research project on tourist-related prostitution in the Caribbean. The support of the ESRC (award no: R000237625) is gratefully acknowledged. I am also indebted to Julia O'Connell Davidson and the anonymous reviewers for their suggestions in revising the paper.

author biography

Jacqueline Sanchez Taylor has undertaken research on sex tourism in the Caribbean, Latin America, India and South Africa. She is currently Lecturer in Sociology at the University of Leeds, UK, where she teaches on the global sex trade and on migration.

references

Agustín, L. (2003) 'Forget victimisation: granting agency to migrants' *Development*, Vol. 46, No. 3: 30–36.

Altman, D. (1999) 'Forward' in P. Aggleton editor, *Men who Sell Sex: International Perspectives on Male Prostitution and HIV/AIDS*, London: UCL Press.

Anderson, P. and Witter, M. (1994) 'Crisis, adjustment and social change: a case study of Jamaica' in E. Le Franc (1994) editor, *Consequences of Structural Adjustment: A Review of the Jamaican Experience*, Kingston: Canoe Press.

Archer-Straw, P. (2000) *Negrophilia: Avant-Garde Paris and Black Culture in the 1920s*, London: Thames and Hudson.

Barriteau, E. (2003) 'Confronting power, theorising gender in the commonwealth Caribbean' in E. Barriteau (2003) editor, *Confronting Power, Theorizing Gender: Interdisciplinary Perspectives in the Caribbean*, Kingston: University of the West Indies Press.

Barry, K. (1995) *The Prostitution of Sexuality*, New York: New York University Press.

Bordo, S. (1993) *Unbearable Weight : Feminism, Western Culture, and the Body*, Berkeley: University of California Press.

Brace, L. (2002) 'The tragedy of the freelance hustler: Hegel, gender and civil society' *Contemporary Political Theory*, Vol. 1, No. 3: 329–348.

Brace, L. and O'Connell Davidson, J. (2000) 'Minding the gap: general and substantive theorizing on power and exploitation' *Signs*, Vol. 25, No. 4: 1045–1050.

Brennan, D. (2004) *What's Love Got to Do with It? Transnational Desires and Sex Tourism in the Dominican Republic*, Durham: Duke University Press.

Cabezas, A. (2004) 'Between love and money: sex, tourism and citizenship in Cuba and the Dominican Republic' *Signs: Journal of Women in Culture and Society*, Vol. 29, No. 4: 987–1015.

Chapkis, W. (1997) *Live Sex Acts*, London: Cassell.

Dahles, H. and Bras, B. 1999 'Entrepreneurs in romance: tourism in Indonesia' *Annals of Tourism Research*, Vol. 26, No. 2: 267–293.

de Alburquerque, K. (1999) 'Sex, beach boys and female tourists in the Caribbean' in B. M. Dank and R. Refinetti (1999) editors, *Sex Work and Sex Workers: Sexuality and Culture*, Vol. 2, New Jersey: New Brunswick.

de Alburquerque, K. (2000) 'In search of the big bamboo: among the sex tourists of the Caribbean' *Transition*, Vol. 8, No. 77: 48–57.

De Moya, A and Garcia, R (1999) 'Three decades of male sex work in Santo Domingo' in P. Aggelton (1999) editor, *Men Who Sell Sex: International Perspectives on Male Prostitution and HIV/AIDS*, London: UCL Press.

de Zalduondon, B. and Bernard, J.M. (1995) 'Meaning and consequences of sexual-economic exchanges: gender, poverty and sexual risk behaviour in urban Haiti' in R.G. Parker and J. Gagnon (1995) editors, *Conceiving Sexuality: Approaches to Sex Research in a Post-modern World*, London: Routledge.

Doezema, J. (1998) 'Forced to choose: beyond the voluntary v. forced prostitution dichotomy' in K. Kempadoo, and J. Doezema (1998) editors, *Global Sex Workers: Rights, Resistance and Redifinition*, London: Routledge.

Dollimore, J. (1991) *Sexual Dissidence: Augustine to Wilde, Freud to Foucault*, Oxford: Clarendon Press.

Dworkin, A. (1979) *Intercourse*, Arrow Books.

Enloe, C. (1989) *Bananas, Beaches and Bases: Making Feminist Sense of International Politics*, London: Pandora.

Fanon, F. (1986) *Black Skin, White Masks*, London: Pluto Press.

Fielding, H. (1996) *The Diary of Bridget Jones*, London: Penguin Books.

Gray, J. (1998) *Men Are from Mars, Women Are from Venus Book of Days: 365 Inspirations to Enrich Your Relationships*, New York: Harper Collins Publishers.

Gunther, A. (1998) 'Sex tourism without sex tourists' in M. Oppermann (1998) editor, *Sex Tourism and Prostitution: Aspects of Leisure, Recreation and Work*, Sydney: Cognizant Communication Corporation.

Hakken, J. (1999) 'Heretical texts: the courage to heal and the incest survivor movement' in S. Lamb (1999) editor, *New Versions of Victims: Feminist Struggle with the Concept*, London: New York University Press.

Herold, E., Garcia, R. and de Moya, A. (2001) 'Female tourists and beach boys: romance and sex tourism' *Annals of Tourism Research*, Vol. 28, No. 4: 978–997.

Hooks, B. (1992) 'Selling hot pussy' in *Black Looks: Race and Representation*, London: Turnaround.

Hooks, B. (1994) *Outlaw Culture: Resisting Representation*, London: Routledge.

Jackson, S. (1997) 'Heterosexuality as a problem for feminist theory' in D. Richardson (1997) editor, *Theorising Heterosexuality: Telling it Straight*, Buckingham: Open University Press.

Jeffreys, S. (1997) *The Idea of Prostitution*, North Melbourne: Spinifex Press.

Jeffreys, S. (2003) *Sex Tourism: Do Women Do It Too?* in *Leisure Studies* Vol. 22, No. 2: 223–238.

Kempadoo, K. (1998) 'Globalising sex workers rights' in K. Kempadoo and J. Doezema (1998) editors, *Global Sex Workers: Rights, Resistance and Redifinition*, London: Routledge.

Kempadoo, K. (1999) editor, *Sun, Sex and Gold: Tourism and sex work in the Caribbean*, Oxford: Rowman and Littlefield.

Kempadoo, K. (2004a) *Sexing the Caribbean: Gender, Race and Sexual Labor*, New York: Routledge.

Kempadoo, K. (2004b) 'Theorising sexual relations in the Caribbean: prostitution and the problem of the 'exotic' in E. Barriteau (2004b) editor, *Confronting Power Theorising Gender: Interdisciplinary Perspectives in the Caribbean*, UWI Press.

Kesler, K. (2002) 'Is a feminist stance of prostitution possible? An exploration of current trends' *Sexuality*, Vol. 5, No. 2: 219–235.

Kitzinger, C. and Wilkinson, S (1993) 'Theorising heterosexuality' in S. Wilkinson and C. Kitzinger (1993) editors, *Heterosexuality: A Feminism & Psychology Reader*, London: Sage publications.

Kleiber, D. and Wilke, M. (1995) *Aids, Sex und Turismus. Ergebnisse einer Befragung deutscher Urlauber und Sextouristen*. Band 33. Schriftenreihe des Bundesministeriums für Gesundheit. Baden-Baden: Nomos Verlagsgesellschaft Baden-Baden.

Lamb, S. (1999) editors, *New Versions of Victims: Feminist Struggle with the Concept*, London: New York University Press.

Lewis, L. (2003) editor, *The Culture of Gender and Sexuality in the Caribbean*, Florida: University Press of Florida.

Lewis, L. (2004) 'Masculinity, the political economy of the body, and patriarchal power in the Caribbean' in B. Bailey and E. Leo-Rhynie (2004) editors, *Gender in the 21st Century: Caribbean Perspectives and Possibilities*, Kingston: Ian Randle Publishers.

MacCannell, D. (1976) *The Tourist: A New Theory of the Leisure Class*, New York: Schocken.

Meisch, L. (1995) 'Gringas and otavalenos: changing tourists relations' *Annals of Tourism Research*, Vol. 22, No. 2: 441–462.

Meyers, N. (2000) director, *What Women Want*, Paramount Pictures.

Mohammed, P. (1998) 'Towards indigenous feminist theorising in the Caribbean' *Feminist Review: Rethinking Caribbean Difference*, No. 59: 6–33.

Mohammed, P. (2003) editor, *Gendered Realities: Essays in Caribbean Feminist Thought*, Kingston: University of the West Indies Press.

Momsen, J. (1994) 'Tourism, gender and development in the Caribbean' in V. Kinnaird and D. Hall (1994) editors, *Tourism: A Gender Analysis*, Chichester: Wiley.

O'Connell Davidson, J. (1998) *Prostitution, Power and Freedom*, Cambridge: Polity.

O'Connell Davidson, J. (2001) 'The sex exploiter' Theme Paper for the Second World Congress Against the Commercial Sexual Exploitation of Children, Yokahama.

O'Connell Davidson, J. and Sanchez Taylor, J (2005) 'Travel and taboo: heterosexual sex tourism to the Caribbean' in L. Schaffner and E. Bernstein (2005) editors, *Regulating Sex: The Politics of Intimacy and Identity*, London: Routledge.

Pateman, C. (1988) *The Sexual Contract*, Cambridge: Polity.

Phillips, J. (1999) 'Tourist-oriented prostitution in Barbados: the case of the beach boy and the white female tourist' in K. Kempadoo (1999) editor, *Sun, Sex and Gold: Tourism and Sex Work in the Caribbean*, Oxford: Rowman and Littlefield.

Phillips, J. (2002) 'The beach boys of Barbados: post-colonial entrepreneurs' in S. Thorbek and B. Pattanaik (2002) editors, *Transnational Prostitution: Changing Global Patterns*, London: Zed Books.

Pruitt, D. and LaFont, S. (1995) 'For love and money: romance tourism in Jamaica' *Annals of Tourism Research*, Vol. 22, No. 2: 422–440.

Sanchez Taylor, J. (2001) 'Dollars are a girls best friend: female tourists' sexual behaviour in the Caribbean' *Sociology*, Vol. 34, No. 3: 749–964.

Scott, S. and Jackson, S. (2000) 'Sexuality' in G. Payne (2000) editor, *Social Division*, London: Macmillan Press.

Seabrook, J. (1996) *Travels in the Skin Trade*, London: Pluto Press,.

Sheller, M. (2003) *Consuming the Caribbean*, London and New York: Routledge.

Sonnet, E. and Whelehan, I. (1995) 'Contemporary identities in women's magazines' in M. Maynard and J. Purvis (1995) editors. *Hetero Sexual Politics*, London: Taylor & Francis.

Williams, P.J. (1997) *Seeing a Colour-Blind Future: The Paradox of Race. The 1997 Reith Lectures*, London: Virago.

Wonders, N. and Michalowski, R. (2001) 'Bodies, borders and sex tourism in a globalised world: a tale of two cities – Amsterdam and Havana' *Social Problems*, Vol. 48, No. 4: 545–571.

World Travel and Tourism Council (WTTC) (2005) The Caribbean: The Impact of Travel and Tourism on Jobs and the Economy, http://www.caribbeanhotels.org/WTTC_Caribbean_Report.pdf.

Yuval-Davis, N. (1997) *Gender and Nation*, London: Sage.

doi:10.1057/palgrave.fr.9400280

83 sexual regimes and migration controls: reproducing the Irish nation-state in transnational contexts

Eithne Luibhéid

abstract

This article examines the ways that state sexual regimes intersect with migration controls to re-make exclusionary nation-states and geopolitical hierarchies among women. I focus on two important Irish Supreme Court rulings: the X case (1992) and the O case (2002), respectively. X was a raped, pregnant, 14-year-old who sought an abortion in Britain. While the Supreme Court ultimately permitted her to procure an abortion, women's right to travel across international borders without government inquiry into their reproductive status came into question. The O case concerned a Nigerian asylum seeker who invoked the fact that she was pregnant in an effort to avoid deportation. The Supreme Court, however, affirmed that she could be deported, despite the Irish Constitution's pledge to protect the 'right to life of the unborn.' Considered together, these cases reveal how overlapping sexual/migration control regimes both reinscribe hierarchies among women based on geopolitical location, and rebound the exclusionary nation-state despite growing transnationalism.

keywords

sexuality; migration; asylum seeker; transnational; Republic of Ireland; X case

(60–78) 0141-7789/06 $30 www.feminist-review.com

introduction

In January 2002, Ms. I.A.O, a 32-year-old Nigerian citizen, made headlines in the Irish Republic (hereafter Ireland) when she appealed to the High Court to prevent her deportation to Nigeria on the ground that she was pregnant. Her solicitor, deploying pro-life rhetoric, claimed that her deportation contravened Article 40.3.3 of the Irish Constitution, which guarantees that the Irish State will 'defend and vindicate' the 'right to life of the unborn.'[1] Ms. O had entered Ireland in December 1999, claiming asylum on the basis that she feared her life would be in danger if she returned to Nigeria. An official who interviewed her judged her claim to be 'manifestly unfounded'[2] and that she had deliberately made false or misleading representations about her case. Ms. O appealed the decision, lost, and was issued with a deportation order. She sought and received permission to institute a judicial review of the deportation order, which did not succeed, and a second deportation order was issued against her on 26 October 2001. Ms. O then sought the services of a firm of private solicitors, who appealed her deportation on several grounds including the fact that she had become pregnant. Building on the Irish Constitution's guarantee of the 'right to life of the unborn,' her solicitor argued that 'deporting her to Nigeria, where the mortality rate was 90 per thousand births, compared with 7 per thousand in the Irish Republic, amounted to jeopardizing the life of the unborn child' (Coulter, 2002).

1 Article 40.3.3 states 'The State acknowledges the right to life of the unborn and, with due regard to the equal right to life of the mother, guarantees in its laws to respect, and, as far as practicable, by its laws to defend and vindicate that right.'

2 This is a legal term that results in reduced scope for appealing a negative judgment.

Other arguments against Ms. O's deportation were also marshalled,[3] but it was her solicitor's deployment of claims concerning 'the unborn' that captured the public imagination. When her case went to the Supreme Court, the *Irish Times*' headline announced, 'Status of the Unborn at the Centre of a Plea Against Deportation' (Carolan, 2002). Unfortunately for Ms. O, the Supreme Court, like the High Court, affirmed that her deportation was legal. In this article, I discuss Ms. O's case in conjunction with the history of Irish women's abortion migration, in order to argue that at the site of the women's bodies, state sexual regimes intersect with migration[4] controls in a manner that reproduces hierarchies among women and exclusionary versions of the nation-state. This occurs despite the growth of transnationalizing processes that have profoundly transformed nation-states. As scholars including Ong (1999) and Westwood and Phizacklea (2000) suggest, transnational processes have reconfigured but not abolished the nation-state's salience in social, economic, and political processes. Similarly, migration scholars observe that even as nation-states open their borders to increasing transnational flows of capital, goods, and information, the control of human migration provides a locus through which nation-states re-bound and 'renationalize' themselves (e.g. Sassen, 1996).

3 These included the arguments that her unborn child was legally a person, and therefore, a deportation order should also have been issued against the unborn child (which it was not); and that Ms. O's asylum claim had not been reviewed according to 'fair procedures' (Carolan, 15 February 2002).

4 Migration includes immigration *and* emigration.

The ways that state sexual regimes figure into these contemporary developments are, however, rarely considered. An analysis of how rebounding of the nation-state frequently operates by subordinating women's sexualities to masculinized, racialized, capitalist nation-states seems to fall outside the purview of much migration (and transnationalism) scholarship. Similarly, while feminist scholarship has elaborated the significance of women's sexualized bodies for constituting the nation-state, there has been limited discussion of the ways that state migration control regimes in the context of growing transnationalization are central to these processes. This article, therefore, focuses on the imbrications of state sexual regulation with migration controls, in a process that results in the reproduction of hierarchies among women and exclusionary nation-states that are positioned within an unequal global order. It asks what feminist politics is capable of addressing these intersections and their consequences.

reproducing the nation-state through gendered sexual regulation

Feminist scholars have thoroughly documented how nation-states construct their identity and sovereignty through subordinating women's sexuality to masculinized nation-building projects (e.g. Anthias and Yuval-Davis, 1989; McClintock, 1993). As Hyunah Yang explains, 'the nation becomes gendered and women's sexuality becomes nationalized. Nation is equated with the male subject position, and women's sexuality is reified as the property of the male nation' (Yang, 1998: 130). Nation-building entails the imposition of heteropatriarchal logic in which women's sexuality is 'perennially colonized within reproduction and heterosexuality' for the masculinized nation-state (Alexander, 1997: 64; see also Berlant, 1997; Smyth, 2005). Women who fail to subordinate their sexuality to heteopatriarchy – including single mothers, lesbians, sex workers, and others – find themselves marginalized within the nation-state, and are often the targets of intervention or criminalization. Heteropatriarchy mandates not only sexual and gender, but also racial and ethnic, hierarchies; as Anthias and Yuval-Davis (1989) explain, women are expected not only to reproduce the collective's members through childbearing and mothering, but to do so in a manner that recreates the boundaries of the racial, ethnic, and/or national collective by having sex with men who are deemed racially/ethnically appropriate. While women of the majority group are exhorted to bear children within heteropatriarchal marriage with 'appropriate' men, racial/ethnic minority women find that their childbearing is often marked as a threat by the state, and controlled, prevented, or demonized accordingly (see Anthias and Yuval-Davis, 1989; Roberts, 1997). Heteropatriarchal marriage also centrally organizes the transmission of property in ways that maintain unequal economic relations (Cott, 2000).

This analysis of the connections between nation-building and female sexual subordination certainly applies in Ireland where, as Ailbhe Smyth argues, 'control of women's sexuality and reproduction has been high on the Irish political agenda since the foundation of the State' (Smyth, 1996: 121).[5] In 1922, after military struggle against British rule, Ireland became partitioned into a Free State (the southern 26 counties) and the North, which remained under British rule. In this postcolonial context, 'the overwhelming push to define Ireland as 'not-England' led to a search for distinguishing marks of identity,' of which women's reproductive sexuality became key (Mullally, 2005: 82). The 1937 Constitution positioned women as mothers and childbearers located within the private home, and this construction involved a claim of moral superiority over the former colonial master. Article 41.2.1 of the Constitution says that 'the State recognises that by her life within the home, woman gives to the State a support without which the common good cannot be achieved.' Consequently, according to Article 41.2.2, 'The State shall ... endeavour to ensure that mothers shall not be obliged by economic necessity to engage in labour to the neglect of their duties in the home.'

5 See also Innes (1993), Meaney (1991), Ryan (2002: 53).

This gender and sexual order, in turn, presumed and institutionalized a specific economic and ethnic/racial logic. A particular vision and version of Irishness as race/ethnicity was enshrined in the Constitution, one that conceived Irish women as settled,[6] Catholic, and 'white.'[7] Recent scholarship draws attention to the fact that 'immigrants, Black Irish, Jewish, Traveller, and Muslim people, among others, are not considered Irish,' or included within most official nationalist representations, including representations of reproduction of the nation through childbearing (Sinha, 1998: 24; see also Lentin, 1998). As we will see in more detail below, these exclusionary representations became the basis on which the nation became materially ordered. Laury Oaks powerfully captures the centrality of reproductive sexuality to Irish nation-building processes: 'in Ireland, [women's] reproduction is a medium through which competing national origin stories that focus on Irish national identity and cultural self-determination, indeed versions of 'Irishness' itself, are imagined and expressed' (Oaks, 1998: 133).

6 The term 'settled' refers to non-Traveller people who are privileged as a result of their settled, non-nomadic status. Travellers are a distinct, racialized group in Ireland, who endure ongoing and systemic discrimination (see O'Connell, 2002).

7 'White' is a term that is generally applied to people with 'light' coloured skin and 'European' features. Yet, whiteness is by no means reducible to these phenotypical characteristics; for example, in the Irish Republic, Travellers generally share these characteristics but remain subjected to racism and are not seen as representative of the nation (see Ní Shúinéar, 2002).

Gaining access to contraception and abortion has been central to women's efforts to assert some degree of control over their own bodies within this heteropatriarchal context.[8] Abortion has been illegal in Ireland since the Offenses Against the Person Act, which passed in 1861 while Ireland was under British rule (Hug, 1999). In 1974, the Irish Supreme Court recognized the right to marital privacy, and anti-abortionists feared that this decision might open the door to the legalization of abortion, as had happened in other countries such as the US (Mullally, 2005).[9] This led to a bitter campaign, and a referendum that resulted in the eighth amendment being added to the Constitution. This is Article 40.3.3, which states 'The State acknowledges the right to life of the unborn and, with due regard to the equal right to life of the mother, guarantees in its laws to respect, and, as far as practicable, by its laws to defend and vindicate that right.' This was the article that was invoked in the Baby O case. That Ireland does not allow abortion has become an important means through which a strongly moral

8 As scholars observe, the fact that pro-life rhetoric is associated with

national identity has been constructed. Mullally affirms: 'the definition of Ireland in exclusively pro-life terms served as another distinguishing mark of Irish identity' (Mullally, 2005: 83).

postcolonialism contributes to the difficulty of challenging it (see Fletcher, 2001).

9 The right to marital privacy was recognized in McGee v. Attorney General [1974] I.R. 284.

Migration controls, as much as sexual regimes, significantly construct the ideological and material boundaries of the nation-state. Indeed, the control of migration is typically viewed as an uncontestable expression of national sovereignty. In Ireland, emigration rather than immigration was the single most important fact of Irish life for nearly two centuries (O'Tuathaigh, 1991: 9). As Mac Laughlin argues, high rates of emigration have 'long been linked to the process of class formation and capitalist development in Ireland' (Mac Laughlin, 1997b: 6), and crucially enabled the formation and consolidation of a nationalist bourgeoisie (Mac Laughlin, 1997b: 13; see also Lee, 1989: 374). After independence, sustained emigration continued. Such high emigration appeared to reflect the failure of the nationalist project to generate a viable livelihood for many, and the government therefore developed strategies for explaining emigration in a manner that exonerated it from responsibility (Lee, 1989: 374; Mac Laughlin, 1997b). Relatively few government interventions, however, were designed to control the outflow of people.[10]

10 For a discussion of state planned migration schemes that were intended to facilitate emigration from Ireland, see Duffy (2004). According to Duffy, after independence, the few government efforts that were intended to facilitate either emigration or relocation within the national territory remained shrouded in secrecy. Note that other groups beside the state, including landlords, philanthropists, and the churches, played important roles in facilitating emigration.

An intimate connection existed between emigration and the nation-state's gender and sexual order. As Tom Inglis describes,

> in the supposed cycle of events after the Famine [of 1845–1849], the segregation of the sexes led to the control of sex. The control of sex helped the control of marriage. The control of marriage helped the control of population and, finally, the control of population helped improve the overall standard of living. ... The strategy of increasing the standard of living through controlling population growth was thus based on restricting the numbers of sons and daughters who could get married.
>
> (Inglis, 1998: 33)

Emigration emerged as the prescribed mechanism for sustaining this sexual order. 'In most families those sons and daughters not selected for marriage had a choice of remaining sexually inactive, or emigrating' (Inglis, 1998: 34). It would be hard to overstate how central emigration remained to the Irish sexual system of the nineteenth and twentieth centuries. As late as the 1960s, an anthropologist studying the Irish could write, 'Young people who anticipate emigration have transferred any expectation of sexual gratification ... to cities far from home. Equally, those who have decided to stay at home do so in the full realization that the decision certainly entails a life of chastity' (cited in Inglis, 1998: 31).

Emigration also emerged as a crucial mechanism for negotiating strict sexual codes, including for women pregnant outside of marriage (Ryan, 2002; Earner-Byrne, 2004), or engaged in same-sex relationships (Luibhéid, 2000), or fleeing incest (O'Carroll, 1995). Single Irish women who were pregnant 'emigrated to

Britain in significant numbers during the twentieth century' (Earner-Byrne, 2004: 155). This form of emigration was widely recognized and critically debated, but the state generally turned a blind eye (Earner-Byrne, 2004: 156). Only in the 1930s, under considerable pressure, did the state actively respond by sanctioning a repatriation programme for pregnant unmarried Irish women in Britain.[11] According to Earner-Byrne, 'the history of this emigration forms a vital part of what has become known as 'the abortion trail'' (Earner-Byrne, 2004: 155).

The abortion trail concerns Irish women facing unwanted pregnancies who have travelled to England and elsewhere for abortions, resulting in a recognizable 'abortion migration' history. Many scholars suggest that abortion migration dates from 1967, when abortion was legalized in Britain. Pauline Jackson, however, argues that Irish abortion migration to England dates from around 1937, when the UK's *Rex v. Bourne* case opened the door to abortion under certain circumstances (Hug, 1999). With the passage of the British Abortion Act of 1967, Irish women's migration to Britain for abortions became more thoroughly institutionalized, even though 'for many women [such travel] is a huge hurdle, involving financial strain, distress and fear, [and] unknown places in an often unknown country' (de Burgh, 2000: 5).[12]

Having to travel to Britain for an abortion drives home for Irish women that they are subordinated citizens within the dominant construction of the nation. Ellen, an Irish woman who had been living in Germany, became pregnant, sought an abortion, and returned to Ireland. She describes:

> I was faced with the fact that I had done something which divided the country. .... In Ireland I became one of those questionable women who'd had an abortion, in Germany I had been a responsible young woman who had dealt with a difficult situation. In Ireland it was something which could hurt others, make others ashamed of me, it was something upon which I would be judged and found wanting. It felt strange to have done something which was so condemned and considered so wrong but which was so right for me.
>
> (Irish Family Planning Association, 2000: 35, 37)

Continuing abortion migration reflects the existence of a state sanctioned sexual regime that subordinates women's bodies to childbearing within patriarchal marriage, while denying access to abortion within the territory of the nation-state. As with the emigration of unmarried mothers, the state has largely turned a blind eye to this form of migration and the social conditions that generate it. In 1992, however, the X case brought these matters into the public spotlight.

X was a 14-year-old Irish girl, pregnant as a result of rape, whose parents brought her to Britain for an abortion. The Attorney General of Ireland issued an injunction against them, and they returned to Ireland without X having procured an abortion. The case was widely covered in the international media, in a manner that generally condemned Ireland. 'The international media reported Ireland to be 'backward,' 'barbarous,' 'punitive,' 'priest-ridden' – a portrayal that did not

11 Earner-Byrne relates that 'repatriation as a tool of reversal or change was utterly unsuccessful because it responded only to the symptoms not the causes of this prenatal emigration' (2004: 172). Causes that Earner-Byrnes (2004: 170) names include extreme social rejection; discrimination; likelihood of being detained in an institution; being held liable for the cost of one's child's maintenance in an institution; being subjected to religious manipulation; and being viewed as a financial and moral burden.

12 Note, 'Irish women have also had abortions in other European countries; in the United States, in Canada, Australia, New Zealand' (Ruane, 2000: 10).

sit well with a modernizing image of an emerging Celtic Tiger economy' (Mullally, 2005: 92). 'Fearing an adverse impact on market forces and the image of Ireland abroad, the government actually volunteered to pay the costs of the Supreme Court appeal, so anxious were they to have the issue resolved' (Murphy-Lawless). Demonstrations in support of X were held in major cities throughout Ireland, as the public was galvanized by her circumstances. In March 1992, the Supreme Court ruled that X could procure an abortion, given that she was suicidal, which represented a threat to her life.[13]

But in the Supreme Court decision, 'the right of a pregnant woman to travel outside the country was contested 3–2 and was not seen as an inalienable right' (Murphy-Lawless; see also Conrad, 2004). Travel by a pregnant woman seeking an abortion 'had never been thought illegal before in Ireland, as [Irish law that criminalizes abortion] does not normally have effect outside the jurisdiction' (Reid, 1992: 26). Three of the five Supreme Court justices ruled in X, however, 'that the implicit right to travel which the Irish Constitution confers on citizens was subordinate to the courts' duty to protect the right to life of the foetus' (Reid, 1992: 26). The possible consequences of this ruling were serious: 'the potential restriction on women's freedom of movement, and with it, the spectre of anti-abortion groups seeking injunctions to restrain pregnant women from travelling abroad, provoked widespread protests' (Mullally, 2005: 93). These included a pro-abortion protest at Dublin airport 'where an "X-ray machine" was used to screen female "passengers" to England. Those found to be pregnant were not allowed to board planes and were "taken into State custody"' (Oaks, 1998: 140). This demonstration dramatized the question of whether and how the state could inquire into women's reproductive status, and adjudicate their migration accordingly. Feminist concerns about women's ability to travel were not unfounded; Angela, who found herself unexpectedly pregnant and who decided to seek an abortion in England, describes that 'those were the days when anti-abortionists were doing everything to stop women from getting to England. I had nightmares that somehow they would stop me before I got to England and force me to continue the pregnancy' (Irish Family Planning Association, 2000: 23). Angela's experience demonstrates the interweaving of sexual and migration controls in the production of the nation-state.

As the X Case unfolded, it emerged that the Irish government had, in 1991, secretly appended a Protocol to the Maastricht Treaty, stating that 'no act or amendment at European level could supersede the 1983 Amendment to the Irish Constitution. In other words, this Protocol denied Irish women recourse to European Community law to overrule the Irish Constitution' on matters of abortion restriction (Murphy-Lawless). 'In June 1992, the Irish electorate had to vote whether to accept a Maastricht Treaty with this Protocol, in the face of all the uncertainty and potential for change that the Supreme Court ruling had opened up' (Murphy-Lawless). Some women's groups urged a 'no' vote, because

13 'The Supreme Court lifted the injunction, reversing the High Court's ruling on the substantive question of abortion. Pointing out the state's duty to have 'due regard' for the life of the mother, the Supreme Court concluded that abortion was lawful in Ireland when there was a 'real and substantial risk' to the life, as distinct to the health, of the mother' (Mullally, 2005: 92).

they argued that Maastricht with the Protocol was not in the best interests of women. Anti-abortion groups also urged a 'no' vote, on the ground that Europeanization allowed EU institutions (especially courts) to interfere in Irish affairs, including on matters of morality and sexuality, in ways that eroded Irish national sovereignty and distinctiveness.[14] As Oaks summarizes, 'the extreme publicity of the interrelationship between abortion politics and the meaning of Irish membership within the European Union created a different space for the presentation of competing sorts of Irishness at a time when 'nation' itself [was] being transformed within the New Europe' (Oaks, 1998: 149).[15]

14 For example, see Oaks (1998).

15 Oaks importantly continues, 'as changes have transformed the meaning of Irishness, so too have 'European' identities been explicitly reformulated within and across the European Union ...' (Oaks, 1998: 149).

Maastricht not only demonstrated the interweaving of abortion politics with national distinctiveness in a context of transnationalization; it was also centrally concerned with nation-state immigration controls. Ironically, the Maastricht Treaty was intended to open up migration among the countries and citizens of the EU – even as the Irish government's Protocol to the Treaty, in conjunction with the uncertainties caused by the Supreme Court decision in X, potentially jeopardized the ability of Irish women of childbearing age to travel across international borders. To facilitate migration among EU countries, the Maastricht Treaty introduced EU citizenship, which 'derived from holding citizenship in a member state and thereby complementing, rather than replacing, national citizenship' (Geddes, 2000: 101).[16] However, while Maastricht opened up migration for EU nationals, it contributed at the same time to further restricting and criminalizing migration by non-EU nationals, including asylum seekers like Ms. O (see Geddes, 2000; Jordan and Düvell, 2004).

16 Contrary to claims about supranational citizenship, this structure reinforced rather than eroded the importance of holding citizenship in a member nation.

The Irish government, eager for the public to ratify Maastricht and concerned that the X case and revelations of the Protocol might derail that process, 'made a Solemn Declaration to try to mitigate the effects of their own Protocol. They promised a separate referendum on abortion travel and information, and the Treaty was passed' (Murphy-Lawless; see also Conrad, 2004: 106–107; Smyth, 2005). In November 1992, the promised referendum on abortion took place, and it involved three different issues. 'Two of the proposed constitutional amendments – guaranteeing pregnant women's freedom to travel abroad and limited rights to the circulation of information about abortion services – were voted into the Constitution in this referendum. A third amendment, proposing a more limited right to abortion than that already allowed by the Supreme Court, was rejected' (Smyth, 1996: 111). Therefore, women who remain subjected to the state's anti-abortion laws at least secured a constitutional guarantee that the state would not prevent them from travelling abroad to procure abortions, should they decide to do this. This guarantee exists side-by-side with the post-Maastricht regime that is intended to facilitate, even generate, migration among EU countries including by Irish citizens.

The events described above show that while anti-abortion policy remains a critical marker of Irish national distinctiveness, the state has been forced to renegotiate the extent and meaning of anti-abortionism in relation to the larger transnational context of EU membership, including the state's desire to encourage ratification of the Maastricht Treaty. Moreover, they show how the state's pro-life stance came to involve questions about the possibility of governmental regulation of Irish women's migration across international borders, based on their reproductive status.

asylum seekers, abortion rhetoric, and Irish national distinctiveness

The troubled intersections of state sexual regimes with migration controls emerged in new form in the late 1990s. At that stage, Ireland's economy was booming, and the state confronted the need to respond to unprecedented levels of immigration into (rather than emigration from) the territory. The Irish government set about developing laws and administrative procedures to manage immigration. These management strategies were viewed as a legitimate expression of national sovereignty.

The construct of national sovereignty, however, and the immigration controls that it authorizes, require critical historical contextualization. As scholarship shows, European sovereign nation-states became consolidated through processes of colonialism and capitalism on a global scale (Castles and Davidson, 2000). The nation form provided a mechanism to extract resources from colonized peoples, and to shield these resources from distribution beyond the boundaries of the designated citizenry. In this sense, the nation state form has been a crucial mechanism through which material inequalities were constructed and maintained on a global scale. Immigration policies that are framed strictly within the horizon of the nation-state, and that thereby fail to address the ways that immigration is driven by unequal global processes, extend the history of the nation form as a mechanism for reproducing inequality (see Mongia, 1999).

For centuries, Ireland remained among the colonized and under-developed regions of the world. After independence, however, it sought integration into the global economy on terms that affirmed, rather than challenged, structural hierarchies (Allen and Loyal, forthcoming). When increasing numbers of immigrants began seeking entry into Ireland in the 1990s, policy makers and the public generally failed to acknowledge that this migration was substantially driven by global dynamics in which Ireland played a distinct role. Instead, immigration was largely framed as a matter of individuals' 'rational choice' decision-making. Within this framework, acceptance of immigrants became a matter of either humanitarianism or else selecting migrants who were deemed 'good' for the nation (on economic, cultural, racial, or other grounds). This reasoning placed immigration control in the service of reproducing the bourgeois, heteropatriarchal, racist nation-state.

Mechanisms for managing immigration significantly pivoted on the creation of administrative categories, to which tiered rights and possibilities were linked. As Allen and Loyal describe, 'the main polarity in the state's categorization of migrants is between asylum seeker/refugee and economic migrant' (Allen and Loyal, forthcoming). According to the 1951 Geneva Convention, asylum seekers are people who are unable or unwilling to return to their countries of origin or habitual residence because they fear persecution on the grounds of race, religion, nationality, membership in a particular social group, or political opinion. The Irish Refugee Act 1996 further states that persecution on the basis of membership in a trade union, gender (being female and male – but not necessarily being transgender), and sexual orientation may provide a basis for gaining asylum. Historically, Ireland was not a major asylum seeker destination. In 1992, just 39 people applied for asylum. But as the numbers grew,[17] 'Ireland aligned itself with current EU policies whose primary [goal] was to reduce the number of "uneconomic" asylum seekers' (Allen and Loyal, forthcoming). Successive redefinitions of who could qualify as an asylum seeker, evident not only in law but also in policy and practice, allowed the Irish state to become more restrictive while appearing to adhere to humanitarian goals. As Liisa Malkki (1995) argues, asylum seekers are not a distinct 'type' of person; rather, governmental processes ultimately determine whether or not someone becomes officially classified as an asylum seeker. As the Irish state sought to limit entry by asylum seekers, it did so not only by trying to pre-empt their arrival in the first place but also by bureaucratic means that invalidated increasing numbers of asylum claims. Bureaucratic invalidation was backed up by a growing discourse that differentiated asylum seekers into 'bogus' versus 'genuine' (Fanning, 2002: 24–25, 101, 105; McGee, 2003: 197), and claims that a majority of asylum seekers were, in fact, 'bogus.' The 'bogus'/'genuine' distinction was often produced by the state's own administrative procedures, rather than by the qualities and experiences of asylum seekers themselves (Garner, 2004). Yet, the state's role in producing that distinction, and the ways that distinction serves an exclusionary nation-state agenda, became lost in public debates and policy decisions.

17 By 2002, nearly 12,000 people applied for asylum in Ireland.

In response to these developments, asylum seekers sought ways to circumvent the state's exclusions. Birthing a child on Irish soil emerged as one critical means to challenge the state's exclusionary regimes, for two reasons. First, until 2005, Ireland offered birthright citizenship, which meant that anyone born on Irish soil was an Irish citizen (Mullally, 2005; Smyth, 2005). Second, in the 1990 *Fajujonu* case, the Supreme Court ruled that Irish citizen children were entitled to the care and company of their parents, within the Irish nation-state, even when these parents were non-nationals.[18] As a result, increasing numbers of asylum seekers abandoned their asylum claims after giving birth in Ireland, and instead

18 For details of this case, see

sought residency based on their parentage of an Irish child.[19] Consequently, asylum seekers' childbearing emerged as extraordinarily controversial. The Minister for Justice and sectors of the media repeatedly claimed that asylum seekers were using childbearing to bypass immigration controls, thereby undermining national sovereignty while multiplying the number of alleged welfare dependents (Luibhéid, 2004). Asylum seeker women in particular became reduced to their sexualized, racialized reproductive organs that, as Nozinic describes, 'they are said to 'use' ... in order to abuse the system' (Nozinic, 2002: 82).

O'Connell and Smyth (2003).

19 Note, in the *L&O* decision, delivered on 23 January 2003, the Supreme Court substantially altered *Fajujonu* in ways that opened the door for the deportation of non-national parents of Irish children.

contesting deportation through the 'right to life of the unborn'

In this context of heated public debate about the intersection of women's sexualities with migration controls, Ms. O filed to prevent deportation. Since she was pregnant but had not given birth, her solicitor argued that the Constitution's commitment to 'defend and vindicate' the 'right to life of the unborn' should invalidate the deportation order against her. However, the government's and courts' responses show how these pro-life discourses and practices became reworked in ways that reproduced and stabilized a distinction between 'bogus' and 'genuine' asylum seekers, thereby harnessing the pregnant bodies of non-national women in service to exclusionary immigration control regimes.

The government's response denied that the Constitutional commitment to 'defend and vindicate' the 'right to life of the unborn' had any application to Ms. O's situation. In the High Court, the State's counsel argued that the woman was not claiming that she faced medical complications associated with the pregnancy, or a specific threat to her life (and therefore to the life of her unborn child) upon return to Nigeria. Rather, counsel argued, 'however dressed up it was, the claim was unfounded. What was being said was that while the claim was being processed, the woman became pregnant and the health service in Ireland was better than the health service in Nigeria (Newman, 2002)'. The High Court concurred that 'in the absence of any adverse medical condition of the mother's, the right to life of the unborn, as enshrined in Article 40.3.3 of the Constitution, was not an issue in the case (Carolan, 2002)'. The Supreme Court affirmed the High Court's judgment:

> The paragraph in Article 40.3.3 on which Counsel relied ... was intended to prevent the legalisation of abortion either by legislation or by judicial decision within the State, except where there was a real and substantial risk to the life of the mother which could only be avoided by the termination of the pregnancy. In this case, neither the State nor any of its organs was seeking to terminate [Ms. O.]'s pregnancy and the fact that the standard of ante- or post-natal care

> available to her in Nigeria was less than would be available to her in this country was entirely irrelevant to the legality of her deportation.
>
> (Supreme Court of Ireland Decisions)

These decisions make clear that the State's commitment to defending and vindicating the right to life of the unborn involves a circumscribed realm: the state will prevent abortion, and possibly intervene if medical complications threaten a pregnancy or if a woman convinces them that she faces a serious threat to her life on being returned to her home country. But otherwise, the State has no responsibility with regard to the life of the unborn. As Mullally writes, 'the threat imposed by higher infant mortality rates could not invoke the protection of Article 40.3.3. The state's duty to defend and vindicate the right to life of the unborn did not extend to ensuring the health and well-being of Baby O, or even ensuring a safe delivery' (Mullally, 2005: 101).

Moreover, even though the mother's circumstances would significantly shape the baby's possibilities in life, the state claimed that it had no further responsibilities toward Ms. O either, since her asylum claim had been duly considered by state representatives and was found to be lacking in merit. Indeed, they judged her to have engaged in offering falsehoods during the asylum process, and denied refugee status to her.

By reframing Ms. O not as a woman deserving of state protection for her unborn child, but as an economic migrant who engaged in falsehoods and attempted to manipulate the asylum system to secure residency in Ireland, the state and courts re-stabilized the state's distinction between a 'genuine' refugee and a 'bogus' asylum seeker who was merely an economic migrant. In other words, their strategy suggested that the distinction between 'genuine' and 'bogus' asylum seekers stemmed not from the government's own shifting methods for adjudicating asylum claims, but from qualities and characteristics of individual migrant bodies, which the government could reliably assess. Their argument, which responded to Ms. O's solicitor's claims regarding 'the unborn,' rendered pregnancy itself as a site through which the 'genuine'/'bogus' distinction could be adjudicated.

Moreover, the state's arguments affirmed the need to police against migrant populations in general, and 'bogus' asylum seekers in particular, while at the same time sidestepping larger issues. For example, as described above, migration flows are substantially driven by longstanding global inequalities. Ifekwunigwe (2004: 406–407) writes in regard to African migrants in Europe that they

> are part of the latest global forced migration system, which in turn grew out of the earlier globalizing phases of transatlantic slavery and Empire (Narayan, 1997). They represent what Catherine Hall describes as a 'reconfiguration of colonial relations' wherein the North and South are still connected by sinews of inequality and subordination.

Migration and asylum policies that fail to substantively address these root inequalities, and instead use migration controls in service to neoliberal capitalist, masculinist, racial states, contribute to maintaining these global hierarchies. The Irish government's creation of a distinction between 'genuine' and 'bogus' asylum seekers – which mirrors the distinctions used in other EU countries – constitutes one crucial mechanism through which these global hierarchies are maintained and legitimized through the migration control system.

The O ruling, which re-stabilized the state's distinction between a 'genuine' and a 'bogus' asylum seeker, inevitably resulted in affirmation of Ms. O's deportation. Deportation, a form of forced migration that is imposed on an individual by the state, remains the ultimate expression of national sovereignty and an important mechanism for re-bordering the nation-state. Ms. O's solicitor highlighted her pregnancy precisely in an effort to prevent her forced migration through deportation. His use of anti-abortion rhetoric was powerful enough to temporarily pause the deportation process. The State and judiciary understood that her claim (like Ms. X's) had implications not just for an individual woman but for all women. Thus, in its submission to the High Court, the State pointed out that if Ms. O's case succeeded, the implication was that 'no woman of childbearing age who could assert that she was pregnant could ever be deported' (Coulter, 2002). Although the High Court rejected Ms. O's appeal, Justice Smyth nonetheless certified

> that his decision involved a point of law of exceptional public importance and that it was desirable in the public interest that an appeal should be taken to [the Supreme Court], the point of law being whether the Minister [for Justice] has the legal right or entitlement to deport a person who had failed to secure a declaration of refugee status from the State because she alleges she is, or is, pregnant.
>
> (Supreme Court of Ireland Decisions)

In the Supreme Court, Ms. O's counsel 'conceded that, if his submission were well-founded, it would necessarily follow that every woman or girl of child bearing age would have to submit to pregnancy testing before she was deported or extradited to a country with less developed pre- and post-natal services than are available in this country' (Supreme Court of Ireland Decisions). In his ruling, Justice Keenan wrote indignantly, 'No such gross violation of the privacy of women and girls could possibly have been intended by the enactment of Article 40.3.3 of the Constitution' (Supreme Court of Ireland Decisions).

Yet, the government clearly had, for some time, been concerned about the nexus between non-national women's pregnancy, migration across international borders, and nation-making. The fact that childbearing enabled some asylum seekers to secure residency heightened the concern. For example, Mr. Jim O'Keeffe, the Justice spokesman for the political party Fine Gael, wondered aloud during parliamentary debates, 'Women who are heavily pregnant are allowed to board aeroplanes. Can we introduce further restrictions in that regard?' Without any apparent sense of irony, the Minister for Justice responded, 'pregnant women have rights too' (O'Halloran, 2004). Yet, just one week earlier, that same Minister published a statement in the *Sunday Independent* regretting that 'to require non-national women of childbearing age to make declarations of pregnancy or otherwise when arriving in the State is clearly unworkable' (McDowell, 2004). It did not seem to occur to these officials that there was anything questionable about governments inquiring into women's reproductive status and circumscribing their mobility accordingly.

conclusion

Taken together, the X and O cases reveal a great deal about the complex inter-relationships between state migration control policies, state sexual regimes, and how national borders – with their resulting hierarchies among women – are continually remade in the context of transnationalism. The two cases connect to larger changes that have been occurring in how Ireland governs migration. Until the 1990s, Ireland was concerned mainly with migration by those who were leaving the country, and paid relatively little attention to those who were entering. The Maastricht Treaty, however, was a crucial element in creating an integrated EU, through which mobile capital and mobile European bodies move with relative ease. The Irish government has eagerly participated in the project of Europeanization, particularly as a means to achieve economic growth, and therefore, it has participated in fostering mobility across the borders and outside the country by its citizens.

At the same time, the mobility of other people, especially asylum seekers, has been systematically de-fostered, disabled, disallowed, and increasingly criminalized. Growing restrictions on asylum seekers' mobility post-Maastricht are complemented by deportation – a form of forced travel that expels them from the states where they have sought asylum. Yet, despite these restrictions, European integration and transnational capital have nonetheless also called into being mobile subjects like Ms. O, for whom new migration possibilities both emerge and become newly necessary, regardless of official European restrictions.

Therefore, reading the X and O cases side by side makes clear that official regimes for governing mobility in the post-Maastricht era position women like Ms. O versus women holding Irish and EU citizenship very differently. This differential positioning is relational, in the sense that mobility for one is thoroughly implicated in growing criminalization and immobilization for the other. Thus, official regimes for governing mobility reiterate national origin as a source of hierarchy and distinct inequality among women – despite transnationalization processes.

These points, of course, also hold true for men. So what is additionally important about the X and O cases is the ways that they highlight women's reproductive sexuality as a critical element in these developments. In both X and O, the question arose as to whether the Irish government could or should inquire into women's pregnancy status, and regulate their mobility accordingly. In both cases, the answer was eventually negative. But that common answer emerged through quite different considerations. With X, concerns about the transnational – especially about the possible impact of negative international opinion on economic development – shaped the final resolution to the case. Other important factors were feminist mobilizations, and concern that the international community might see Ireland as lesser than, rather than equal to, the rich and powerful nations.

In the O case, it was also decided that the government should not inquire into women's reproductive status and regulate their mobility accordingly. But that decision emerged through a very different set of concerns. Unlike in X's case, Ms. O's solicitors actually argued that her mobility *should* be mediated through government consideration of her pregnancy. This troubling claim was couched in the powerful language of the pro-life state, and ironically, if accepted, the result would have been *positive*, not only for Ms. O but also for potentially thousands of migrant women. The government's rejection of Ms. O's claim was couched in terms of a very narrow reading of what constitutes a pro-life position, and also in terms respecting women's 'privacy,' and these in turn were harnessed to discourses and practices of immigration and asylum law. The impact for Ms. O and for migrant women more generally was to foreclose opportunities and possibilities for legalizing their presence in the state, or for securing opportunities for their children.

Through these processes, state reproductive politics directed toward controlling the sexualities of Irish citizen women became clearly differentiated from state politics concerning the sexual control of non-national women, especially asylum seekers. Through women's reproductive sexuality, the state re-drew and re-stabilized the line between national and non-national women, and, within the latter group, the line between 'bogus' and 'genuine' asylum seekers. Effectively, the state turned women's sexualities into a tool for racialized strategies of immigration control, which rebounded the Irish nation-state in the context of transnationalism.

Ms. O's case demonstrates that the pregnant bodies of non-national women provide particularly powerful loci through which the state both extends and legitimizes its exclusionary immigration practices. Since Ms. O's deportation, the invocation of pregnant asylum seeker women's bodies to extend and legitimize exclusion has intensified. For example, in January 2003, in *Lobe and Osayande v. Minister for Justice*, the Supreme Court effectively overturned the *Fajujonu* decision by ruling that the Minister for Justice could deport non-national parents of Irish children in the interests of 'the common good,' including the state's right to control immigration (see O'Connell and Smyth, 2003). In June 2004, after alarmist accounts about pregnant migrant women who allegedly engaged in 'citizenship tourism' by travelling to give birth to children on Irish soil, the Irish electorate overwhelmingly approved a referendum that ended the practice of birthright citizenship. In these instances, non-national women's sexualities occasioned the re-bounding of the nation-state and the tightening of exclusionary immigration laws.

What feminist sexual politics would be capable of linking the exclusion of pregnant asylum seeker women to the sexual subordination of Irish women (as reflected by their continuing abortion migration), and challenging both? Until such politics develops, Irish and EU feminists remain implicated in the state's ongoing differentiation and hierarchization of women by nationality, and the resulting reproduction of heteropatriarchal, racialized, economically exploitative nation-states in the context of global inequality.

acknowledgements

Research funding was generously provided by the Wenner Gren Foundation for Anthropological Research. Warmest thanks to Hai Ren, Rob Buffington, Ronit Lentin, Breda Gray, and two anonymous reviewers for feedback and suggestions. I am responsible for any errors.

author biography

Eithne Luibhéid is the Director of LGBT Studies and an Associate Professor of Women's Studies at the University of Arizona, USA. She is the author of *Entry Denied: Controlling Sexuality at the Border*, University of Minnesota Press, Minnesota (2002), and co-editor of *Queer Migrations: Sexuality, US Citizenship, and Border Crossings*, University of Minnesota Press, Minneapolis (2005). Her research focuses on the intersections of sexual regulation, immigration control, and racialization processes in transnational contexts.

references

Alexander, M.J. (1997) 'Erotic Autonomy as a Politics of Decolonization: an Anatomy of Feminist and State Practices in the Bahamas Tourist Economy' in Alexander M.J. and Mohanty C.T. (1997) editors, *Feminist Genealogies, Colonial Legacies, Democratic Futures*, New York: Routledge, 63–100.

Allen, K. and Loyal, S. (forthcoming) 'Rethinking immigration and the state in Ireland' in Lentin, A and Lentin R. (forthcoming) editors, *Race and State*, Cambridge: Cambridge Scholars Press.

Anthias, F. and Yuval-Davis, N. (1989) 'Introduction' in Yuval-Davis, N. and Anthias, F. (1989) editors, *Woman-Nation-State*, New York: St Martins, 1–15.

Berlant, L. (1997) *The Queen of America Goes to Washington City*, Durham: Duke University Press.

Carolan, M. (2002) 'Status of the unborn at the centre of a plea against deportation' *Irish Times* 6 February.

Carolan, M (2002) 'Court appeal against deportation fails' *Irish Times* 15 February.

Castles, S. and Davidson, A. (2000) *Citizenship and Migration: Globalization and the Politics of Belonging*, New York: Routledge.

Conrad, K. (2004) *Locked in the Family Cell*, Madison: University of Wisconsin Press.

Cott, N. (2000) *Public Vows. A History of Marriage and the Nation*, Cambridge, MA: Harvard University Press.

Coulter, C. (2002) 'AG to argue that unborn child not a person' *Irish Times* 9 January.

de Burgh, S. (2000) 'Preface' in Irish Family Planning Association (2000) *The Irish Journey. Women's Stories of Abortion*, Dublin: Irish Family Planning Association, 4–5.

Duffy, P. (2004) editor, *To and from Ireland: Planned Migration Schemes c. 1600–2000*, Dublin: Geography Publications.

Earner-Byrne, L. (2004) 'Moral repatriation: the response to Irish unmarried mothers in Britain 1920s–1960s' in Duffy P. (2004) editor, *To and from Ireland: Planned Migration Schemes c. 1600–2000*, Dublin: Geography Publications, 155–173.

Fanning, B. (2002) *Racism and Social Change in the Republic of Ireland*, Manchester: Manchester University Press.

Fletcher, R. (2001) 'Post-colonial fragments: representations of abortion in Irish law and politics' *Journal of Law and Society*, Vol. 28, No. 4: 568–589.

Garner, S. (2004) *Racism in the Irish Experience*, London: Pluto Press.

Geddes, A. (2000) *Immigration and European Integration. Towards Fortress Europe?*, Manchester: Manchester University Press.

Hug, C. (1999) *The Politics of Sexual Morality in Ireland*, New York: St Martins.

Ifekwunigwe, J. (2004) 'Recasting 'Black Venus' in the new African diaspora' *Women's Studies International Forum*, Vol. 27, No. 4: 397–412.

Innes, C.L. (1993) *Woman and Nation in Irish Literature and Society*, Athens: University of Georgia Press.

Inglis, T. (1998) *Lessions in Irish Sexuality*, Dublin: University College Dublin Press.

Irish Family Planning Association (2000) *The Irish Journey. Women's Stories of Abortion*, Dublin: Irish Family Planning Association.

Jordan, B. and Düvell, F. (2004) *Irregular Migration*, Cheltenham: Edward Elger.

Lee, J.J. (1989) *Ireland 1912–1985. Politics and Society*, Cambridge: Cambridge University Press.

Lentin, R. (1998) 'Racializing (our) Dark Rosaleen: Feminism, racism, antisemitism' *UCG Centre for Women's Studies Review*, Vol. 6: 1–18.

Luibhéid, E. (2000) 'The pink tide: narrating Ireland's lesbian and gay migrations' *Journal of Commonwealth and Postcolonial Studies*, Vol. 7, No 1: 149–168.

Luibhéid, E. (2004) 'Childbearing against the State? Asylum seeker women in the Irish Republic' *Women's Studies International Forum*, Vol. 27, No. 4: 335–350.

Mac Laughlin, J. (1997b) 'Emigration and the construction of nationalist hegemony in Ireland' in Mac Laughlin, J (1997) editor, *Location and Dislocation in Irish Society: Emigration and Irish Identities*, Cork: Cork University Press, 5–35.

Malkki, L. (1995) 'Refugees and exile: from 'refugee studies' to the national order of things' *Annual Review of Anthropology*, Vol. 24, No. 1: 495–523.

McClintock, A. (1993) 'Family feuds: gender, nationalism, and the family' *Feminist Review*, Vol. 44: 61–80.

McDowell, M. (2004) 'Proposed citizenship referendum' *Sunday Independent* 14 March.

McGee, H. (2003) 'Media response to asylum in Ireland' in Fraser U. and Harvey C. (2003) editors, *Sanctuary in Ireland. Perspectives on Asylum Law and Policy*, Dublin: Institute of Public Administration, 181–202.

Meaney, G. (1991) *Sex and Nation: Women in Irish Culture*, Dublin: Attic Press.

Mongia, R. (1999) 'Race, nationality, mobility: a history of the passport' *Public Culture*, Vol. 11, No. 3: 527–556.

Mullally, S. (2005) 'Debating reproductive rights in Ireland' *Human Rights Quarterly*, Vol. 27, No. 1: 78–104.

Murphy-Lawless, J. (2001) Fertility, bodies and politics: the Irish case,' at http://www.hsph.harvard.edu/Organizations/healthnet/reprorights/docs/murphy.html,accessed 18 July.

Narayan, U. (1997) *Dislocating Cultures: Identities, Tradition and Third World Feminism*, London: Routledge.

Newman, C. (2002) 'Counsel says woman's previous claims were unfounded' *Irish Times* 10 January.

Ní Shúinéar, S. (2002) 'Othering the Irish (Travellers)' in Lentin R. and McVeigh R. (2002) editors, *Racism and Anti-Racism in Ireland*, Belfast: Beyond the Pale Publications, 177–192.

Nozinic, D. (2002) 'One refugee experience in Ireland' in Lentin R. and McVeigh R. (2002) editors, *Racism and Anti-Racism in Ireland*, Belfast: Beyond the Pale Publications, 73–90.

O'Connell, D. and Smyth, C. (2003) 'Citizenship and the Irish Constitution' in Fraser U. and Harvey C. (2003) editors, *Sanctuary in Ireland. Perspectives on Asylum Law and Policy*, Dublin: Institute of Public Administration, 229–245.

O'Connell, J (2002) 'Travellers in Ireland: an examination of discrimination and racism' in Lentin R. and McVeigh R. (2002) editors, *Racism and Anti-Racism in Ireland*, Belfast: Beyond the Pale Publications, 49–62.

O'Carroll, I (1995) 'Breaking the silence from a distance: Irish women speak on sexual abuse' in O'Sullivan P (1995) editor, *Irish Women and Irish Migration*, Leicester: Leicester University Press, 192–200.

O'Halloran, M. (2004) 'Postpone citizenship referendum, urges FG' *Irish Times* 22 April.

O'Tuathaigh, M.A.G. (1991) 'The historical pattern of Irish emigration: some labour aspects' in Galway Labour History Group (1991) editor, *The Emigrant Experience*, Galway: Galway Labour History Group, 9–28.

Oaks, L. (1998) 'Irishness, Eurocitizens, and reproductive rights' in Franklin S. and Ragoné H. (1998) editors, *Reproducing Reproduction: Kinship, Power and Technological Innovation*, Philadelphia: University of Pennsylvania Press, 132–155.

Ong, A. (1999) *Flexible Citizenship*, Durham: Duke University Press.

Reid, M. (1992) 'Abortion Law in Ireland after the Maastricht Referendum' in Smyth A. (1992) editor, *The Abortion Papers Ireland*, Dublin: Attic Press, 25–39.

Roberts, D. (1997) *Killing the Black Body*, New York: Vintage.

Ruane, M. (2000) 'Introduction' in Irish Family Planning Association (2000) editor, *The Irish Journey. Women's Stories of Abortion*, Dublin: Irish Family Planning Association, 6–11.

Ryan, L. (2002) 'Sexualising emigration: discourses of Irish female emigration in the 1930s' *Women's Studies International Forum*, Vol. 25, No. 1: 51–65.

Sassen, S. (1996) *Losing Control? Sovereignty in an Age of Globalization*, New York: Columbia University Press.

Sinha, S. (1998) 'The right to Irishness: implications of ethnicity, nation and state towards a truly multi-cultural Ireland' in Lentin R. (1998) editor, *The Expanding Nation: Towards a Multi-Ethnic Ireland*, Dublin: Trinity College Dublin.

Smyth, A. (1996) 'And nobody was any the wiser: Irish abortion rights and the European union' in Elman R.A. (1996) editor, *Sexual Politics and the European Union*, Providence and Oxford: Berghahn Books, 109–130.

Smyth, L. (2005) *Abortion and Nation. The Politics of Reproduction in Contemporary Ireland*, Aldershot: Ashgate Publishing Limited.

Supreme Court of Ireland Decisions, Baby O. v. Minister for Justice, Equality and Law Reform [2002] IESC 44 (6 June, 2002).

Westwood, S. and Phizacklea, A. (2000) *Trans-nationalism and the Politics of Belonging*, London: Routledge.

Yang, H. (1998) 'Remembering the Korean military comfort women. Nationalism, sexuality, and silencing' in Kim E.H. and Choi C. (1998) editors, *Dangerous Women. Gender and Korean Nationalism*, New York: Routledge, 123–139.

doi:10.1057/palgrave.fr.9400281

83 moral ambivalence and irregular practices: contextualizing male-to-male sexualities in Calcutta/India

Paul Boyce

abstract

Male-to-male sexuality in India has been described as both heavily stigmatized and implicitly tolerated. This paper examines these apparently contradictory attitudes, arguing that they reflect broader moral ambivalence about homosexuality in Indian culture and society. While the effects of homophobia in India are very real, simultaneous social latitude allows for relatively un-scrutinized same-sex sexual contact. The paper explores this scenario as a post-colonial legacy and considers the consequences for contemporary sexual subjectivity, particularly in respect of irregular responses to emerging gay identities and socially ambiguous male-to-male sexualities. Conceiving of men who have sex with men as subject to both prejudice and tolerance raises complex questions for HIV/AIDS related policy, programming and activism. The paper argues that understanding male-to-male sexualities in India as practiced within a climate of ambiguous moral censure offers critical insights for the future promotion of health.

keywords

sexuality; morality; homophobia; colonialism; India; HIV/AIDS

(79–98) 0141-7789/06 $30 www.feminist-review.com

introduction

A recent web-based BBC news article entitled 'Fear and loathing in gay India' (http://news.bbc.co.uk/2/hi/south_asia/4304081.stm) reviews shifting attitudes toward gay and lesbian identity and relationships in the sub-continent. Accounts of a movement towards 'greater acceptance' are countered with sobering tales of intransigent prejudice. A report of the growth of a gay, lesbian and transsexual march in Calcutta (attended by 300 people in 2004) is paralleled by continuing testimonies of emotional isolation and social ostracism.

The web-pages include comments on the article posted by various contributors from around the world. The two responses excerpted below reveal differing beliefs about prejudice directed at gay men and other men who have sex with men in India.

Quote 1:

> Being an Indian I understand the culture and the various restrictions there are. I face up to a lot in my life with my parents and relationships outside. In my opinion people should do what they feel is right for themselves. However this causes so many problems with families who end up breaking apart. People should accept each other for who they are which shows development in understanding. I don't exactly agree with gay relationships but I wouldn't judge anyone because everyone is different and the sooner that every one accepts this the happier people will be. Places like India are too culturally bound and I feel they need to expand their horizons and not be afraid of true reality.
>
> Gurjit, England

Quote 2:

> Having spent some time in Calcutta last year, I experienced many encounters with other gay men. Although I was not looking for a partner at the time I was approached quite openly and propositioned in the street. My status as a Westerner no doubt contributed to this, but even clubbing with Calcuttan friends I was treated just the same as my straight friends, if with a little more interest. I think there is no doubt that homosexuality is a huge issue in India, but it's a Western desire to assert one's identity as one thing or another that is upsetting the balance. There is no Gujarati word for 'gay', for instance, and I had a relationship with a man in Delhi who though dismissed by his colleagues as 'not normal' was nonetheless accepted, along with his open sexuality.
>
> Charlie, England

The first quote depicts a conservative India, bound by familial traditions and circumscribed by fear of change. This articulates closely with a wider position, in which Indian culture has been conceived as generally repressive of individual sexual expression and especially women's sexuality (Kartrak, 1992; Uberoi, 1996; Vanita 2002a). In this climate 'gay relationships', according to the commentator, have little or no place, since they disrupt conservative mores and family values.

The existence of same-sex sexuality in India has often been contested, and analyses and popular belief have tended to fixate on the nominally 'third-gendered' figure of the *hijra* as a cultural repository of non-heteronormative sexuality (Cohen, 1995, 2005). This is an outcome of colonial era European fascination with 'third sex,' which became an especially popular trope in post 1870s ethnography (Bleys, 1996: 2). Alternatively studies of (same-sex) sexuality in South Asia has been restricted to historiography (Srivastava, 2004), especially with respect to the putative emasculation of Indian men in the context of British colonialism, and related beliefs about homosexuality and moral degeneracy (Sinha, 1997; Vanita and Kidwai, 2000). The other principal context for debate about male-to-male sexuality in India is psychoanalytic. Theory in this area has generally conceived male-to-male subject–object relations as indicative of sublimated homosexuality in the constitution of 'Indian masculinity' (Kakar, 1997, 2002; Osella *et al.*, 2004).

While resonant of broader social contexts, the perspectives summarized above occlude understanding of male-to-male sexuality as an actual *lived* experience for contemporary men in India. This analytical absence arguably conspires with the lack of social acceptability and legitimacy afforded such relationships in contemporary Indian culture. As Gayle Rubin (1999) has argued, socially censured sexualities are rarely afforded analytical complexity. Against this background, Gurjit (in quote one above) would like to see a more open, tolerant society, accepting of 'gay relationships' (in his terms). Despite this, however, Gurjit admits that he himself does not 'exactly agree with gay relationships,' an attitude that reflects the values he simultaneously questions. This indicates something of an ambivalent moral position, private tolerance coupled with public reprobation – an attitude arguably typical of wider contemporary social values regarding same-sex sexuality in India.

In the second quote above Charlie puts forward a different, although similarly ambiguous point of view. He describes a scenario in which tolerance is more common than conventional accounts of cultural conservatism might suggest. Acceptance in bars and open cruising on the streets of Calcutta hardly indicate a furtive response to sexual repression. This comment can be read in reference to a tradition of travelogue writing by foreigners who have recounted liberal attitudes towards male homosexuality in India. E.M. Forster, for example, wrote of acceptance of same-sex affairs he had whilst working for the Maharaja of Dewas in 1921 (Vanita and Kidwai, 2000). Contemporary Indian authors (and activists) have also written on this theme. For instance Ashok Row Kavi (1999) has told of the moral relativism offered to him by a Swami at a Ramakrishna Mission, who allayed his anxieties that homosexuality was 'wrong'. In a similar vein, Charlie's account of his lover's status among his peers indicates social acceptance. Yet, at the same time this moderate view is shot through with an implicit counter discourse. The lover in question was dismissed as 'not normal' by his supposedly open-minded colleagues. This suggests conflicting attitudes, indicative of tolerance and disapproval as parallel processes.

Against this backcloth, this paper aims to explore ambivalent attitudes and ambiguous moral censure towards male-to-male sexuality in contemporary India. As indicated above, the Indian social context is one in which repressive attitudes and social latitude towards same-sex sexuality appears to be particularly intertwined. This paper examines some of the kinds of sexualities that take shape between men in this context. Ambiguous prejudice arguably produces ambiguous sexual subjects.

Although in part the paper is concerned with the experiences of men who are conceiving subjectivities in response to (nascent) understandings of same-sex sexual identity, this is not to insist on viewing male-to-male sexuality in India through the 'lens' of Western gay identification. The paper is derived from fieldwork that was largely concerned with the experiences of men for whom same-sex sexual acts bore little or no linear correspondence to a sense of sexual subjectivity. Nevertheless, some men were, and are, concerned with understanding self in respect of same-sex sexual identification. Superficially the subjectivities taking shape in this context may look like identity-oriented forms of same-sex sexuality, but they are far from straightforwardly so.

Demographically in Calcutta, as elsewhere in India, men who identify according to same-sex sexuality constitute a minority compared to the larger population of men who have sex with men. Politically, however, men who identify are more vocal in local and national activism. Gay, lesbian and queer activists have been significant in contemporary praxis in India – bringing together conceptual views on South Asian sexualities with rights-based political projects (Vanita, 2002a). This is not to disregard the significance of activists who have taken agendas forward in terms of putatively indigenous terms of identity. Nor is it to ignore the political significance of non-identified same-sex sexual subjects (Boyce, forthcoming). Indeed, the rights of such people in India figure increasingly in activists' debates, raising pressing questions about the terms in which 'non-identified' sexual constituencies are included in sexual politics. Akshay Khanna (2005), for example, has asked why same sex sexual subjects (in India) should feel compelled to define themselves in terms of who they have sex with. What is the relevance of such identity oriented same-sex sexualities to India – a cultural context in which the social scope of such thinking is far from common and generally restricted to urban social milieu?

Same-sex sexual identities in India can be read as the product of both colonial legacies and contemporary Western-oriented identity matrices, such that, as Khanna observes 'the options we have, to understand and address our realities are regulated by our colonial histories and our imperialist presents' (Khanna, 2005: 91). Given this, meanings evoked by people employing terms of same-sex sexual identity are especially interesting, as they evince complex responses to past and contemporary emphases on sexual classification, enacted in a context in which the cultural power of such categories is nebulous. Resistance to sexual identification is often unspoken – or not conceived as resistance at all, such that even men and women who use categories of same-sex sexual identity may simultaneously align to non-identity oriented conceptions of sexuality. This paper explores these partial practices, raising questions about the coherence of sexuality as a discretely identifiable realm of experience.

The paper draws on recent ethnographic work with men who have sex with men in Calcutta, which in turn drew on the author's prior (and ongoing) experience of working in HIV prevention for men who have sex with men in the city (and elsewhere in South Asia). Fieldwork was carried out for 18 months from May 2000 to October 2001. While consisting of interviews, participant observation and group discussions, the study simultaneously blurred boundaries between 'research' and 'life', which was vital to an understanding of empirically intangible aspects of sexuality. The ethnography particularly explored relationships between sexual subjectivity and embodiment, spatialization of sexual practice in the city and contested understandings of kinship in the context of emerging gay identities. Findings support a critical analysis of HIV/AIDS policy and programming aimed at addressing the rights and needs of men who have sex with men in India, advocating the necessity of conceiving non-identified same-sex sexual subjects as intrinsic to successful health promotion and activism.

The following section presents ethnographic material that complicates uniform interpretations of gay identification, illustrating ways in which men who have sex with men in Calcutta take up gay terminology while simultaneously demurring at the individuated same-sex sexual identity implied. This sets the scene for a broader examination of less identity-oriented expressions of sexuality. The next section contextualizes elements of the contemporary ethnography, examining the moral context of same-sex sexuality in India, as a post-colonial legacy of British imperial homophobic discourse and policy (Vanita and Kidwai, 2000; Kugle, 2002). This introduced values premised on Westernized ideologies of individuated sexual subjects to India. However, simultaneous privileging of family values and arranged marriage in the sub-continent means that same-sex sexual relationships have rarely been radically disassociated from heterosexuality – since obligation to marry prevails (Vanita, 2002a). As such the moral forces and diffuse technologies of power (in Foucauldian terms) that have helped to define the homosexual as a discrete person in nominally Western contexts have different effects in India, wherein homophobia can be interpreted as *simultaneously strong and weak* – with irregular outcomes for sexual subjectivity and practice.

The implications of this ambiguous moral scenario are considered by exploring ways in which men who have sex with men in Calcutta (and elsewhere in India) carry out relationships through putatively heteronormative social connections.

The concluding section discusses sexual alterity within ostensibly heterosexual kinship and other social relations, considering the potential for conceiving 'non-identified' subjects in analysis and activism. This is especially relevant to HIV prevention work with men who have sex with men in India, in which context denial that such men exist has been a common barrier to advocacy, programme funding and supportive policy. Moreover, conceiving men who have sex with men as ambiguous sexual subjects is vital to successful health promotion in this context.

displacing 'gay': ambivalent positions among men who have sex with men in Calcutta

In the quotes from the BBC article included above, the respondents both use the term 'gay'. Whether this is intended to specifically refer to self-identified gay men, or men who have sex with men in a more collective way is a little unclear. Certainly, in the first quote Gurjit appears to use 'gay' as a generic term. This is arguably indicative of ways in which sexual subjectivity and identity are commonly conflated in popular (nominally Western influenced) discourses.

The second quote above makes some differentiation, specifically citing the emergence of 'gay' and putatively Western beliefs about identity as responsible for 'upsetting the balance,' introducing individuated expressions of sexual identity. This implies a previous and parallel 'non-gay' context of same-sex sexuality. The disturbance premised in the quote reflects popular attitudes about the nature of identity in India (dividuated rather than individual), and specifically a view that 'gay' is inherently Western and alien – affecting indigenous sexual cultures.

The promotion of gay identity in India has been related to contexts of global and Western acculturation, associated with urban club scenes, etc. (Reddy, 2004). Yet this is only one perspective, and the subtle and diffuse ways in which men in India (as elsewhere) are appropriating gay identity are far from uniform, and are infused with locally conceived meaning (Cohen, 2002; Vanita, 2002b; Reddy, 2004). Moreover as Ruth Vanita (2002a) has argued, asserting gay as unequivocally Western denies the two hundred year history of English as an Indian language, and the contemporary evolution of Indian English in respect of 'modern' sentiments and sexual subjectivities. 'Gay' can be, and is, Indian, even if, or because, it refers to Western ideas, language and influences.

As much as 'gay' may be Indian, it does not necessarily work as a clear statement of a distinctly 'modern' sexual identity, in contradistinction to other putatively 'indigenous' discourses of less socially defined male-to-male sexuality. To take an example from my Calcutta ethnography, Shusham[1] was someone whom I would meet on a semi-regular basis when spending time in a popular cruising area in the south of the city. In response to a research question asking people to reflect

1 All names used in this paper are pseudonyms and identifying details of the people featured have been disguised and 'fictionalized' so as to respect anonymity.

on the first time they heard a word that they felt described their sexual feelings, Shusham recounted how once, when he was at boarding school, he was perusing an English dictionary and came across the word 'gay'. Shusham remembered this as a profound epiphany, a sudden correspondence between his own nascent sexual feelings towards other boys and a word that made some sense of his conflicting emotions.

Another respondent, Kantik, told of how he first happened upon the word 'gay' when reading a report in one of Calcutta's daily English newspapers about the Hollywood actor Rock Hudson's infection with HIV, which was widely reported in the global media in the early 1980s. As with Shusham, this was described as a moment of recognition of self. These two relatively simple examples echo a common trope in the typical Western 'coming out' story, which commonly starts with 'trips to the public library, furtive searches through dictionaries, or secret reading of novels that explore lesbian and gay topics' (Donham, 2005: 273 in reference to Newton, 1984). Despite the common narrative, however, I want to emphasize that neither Shusham nor Kantik identified the word gay as involving an expansion of their sense of self to include a Westernized subjectivity, despite the obvious connections to English as a medium – and in the latter case, the conveyance of 'gay' through a story about a Hollywood actor. 'Gay' was a word that they took as their own, as it was made available to them within their own milieu.

Conceiving 'gay' as having multiply located origins complicates interpretation of trans-national influences on male-to-male sexuality. As Donald Donham (2005) has argued, although ethnographies that explore how gay identification is being transmitted across space (from, say, North to South) may be significant, it is also necessary to understand how such an apparently global identity matrix operates within the local context. Donham recognizes that male-to-male sexuality in Soweto (his study site) has not been so much modernized as layered with new meanings. This observation signifies a broader need for ethnographies that explore gay and lesbian praxis that is not recognizably Western, cognizant of nuanced and interstitial practices. I would add that this not only means tracing how sexual 'liberation' may be imagined locally, but also how practices such as gay acculturation may have little to do with claiming explicit identities or rights and may not be neatly aligned to any kind of uniform, globally intelligible liberatory narrative, conceived as identity oriented reactions to prejudice. Silence and/or indeterminacy in the subject positions people occupy with respect to homosexual and heterosexual differentiations are equally salient. This may be especially so in a context such as India in which the moral boundaries that have helped to generate explicit gay identities in the West have not taken shape in such a way as to produce a culture in which strongly identified gay subjects are generically salient.

During the time of my fieldwork, the main community support group for gay men in Calcutta was called Counsel Club (henceforth 'CC', a popular abbreviation used by the group's members). The leaders and founder members of CC described their social and political agenda as explicitly advancing a gay rights manifesto and offering social and emotional support to gay men in the city. The organization cultivated associations with international lesbian and gay groups, and articulated a programme aligned to Western identity politics. Yet CC was far from didactic in asserting gay identity over its members. Although the principal organizers of the group predominantly identified themselves as gay to friends, family and sometimes within the media, they recognized that many of their membership occupied more interstitial positions with respect to sexual identification. Many members of CC referred to themselves as gay in casual conversation, but were far from wedded to explicit gay identification. This could produce, what – from a gay liberatory praxis reading – looks like compromised statements of emancipation. For example, Goutam, one respondent spoke in the following terms:

> Counsel Club is an organisation for gays, lesbians, bisexual men and women and transgender and transsexual people like us, which provides to these people. Like emotional support, friendly support and several other kinds of support as well. Therefore here we get a space where we can talk freely. So earlier I did not have the courage that I could tell people on their face, but after I came here I could tell my friends some things about this issue. Like the other day I asked one of my friends his opinion about gay people. He replied, 'No, that's a disease, it has been proved to be a disease in America.' Then I told him that it had been proved that homosexuality is no longer a disease. When I told him this quite strongly he ultimately believed that it's not a disease and then he said, 'OK, it could be normal,' but he would object if two gay guys pass comment or tease some ordinary guy. Otherwise he does not have any problem. What I want to say from this is that now I can more or less talk freely about these things, especially because he [the friend] does not know about my sexuality.

This account is hardly a blow for 'out' gay liberation, given that the premise of the respondent's articulation of empowerment is founded on secrecy about his sexual orientation. This is not to disparage Goutam's stance. What he describes is a personal choice. Moreover, his way of addressing his friend's attitudes took place within a context in which talk about one's own homosexuality would, in the contemporaneous general Bengali/Indian social context, be strongly disfavoured, cause for exposure to bigotry, or indeed (and perhaps more pertinently) blank incomprehension.[2] Some members of CC who were familiar with the notion of 'coming out' expressed horror at the thought of ever exposing themselves in such a way. For others, however, the whole idea of identifying oneself according to sexual object choice (as for example gay or homosexual) appeared cognitively dissonant. For many members of CC, sexuality was not based on explicit same sex subject–object relations and an ensuing identification of gay selfhood derived

2 Neither is this a culturally peculiar phenomenon. Prototypically Western 'coming out' narratives offer accounts of usually young gay, lesbian and/or queer subjects as they navigate self-revelation and concealment. Recent work with young gay men and lesbians in the UK has reiterated the navigation of secrecy, exposure and the choice to 'come out' as a persistent rite of passage, rather than a straightforward coming out trajectory (Warwick *et al.*, 2003).

from such an understanding. Importantly, this did not rule out the use of the word gay as an epithet for identification.

I recall an incident when an American businessman, Leo, who had recently been posted to Calcutta, was keen to meet some fellow gay men in the city and so contacted CC to arrange a meeting. Leo organized a small gathering at his house and my friend, the director of CC, invited me along. During the course of the evening, Leo was especially interested to find out more about the lives of gay men in the city and he asked two men at the gathering, who had introduced themselves as a gay couple, to reflect on their families' attitudes towards their relationship. One of the men, Bishwas, simply answered that he did not discuss these matters with his parents and hence it was not a manifest issue. His partner, Deb, on the other hand replied, without any apparent sense of contradiction, 'Oh, but I'm heterosexual!' Under some inquisitive probing from Leo, it became apparent that Deb was not suggesting that he was 'hetero' with his parents and family and 'homo' with his CC friends and lover. Such dualities were salient and certainly the presentation of himself as heterosexual to his family did affect secrecy about a same-sex sexual life that he could never imagine discussing with them. In a sense, this was a default presentation of self. It was not that Deb actively announced to his parents that he was heterosexual. Rather this was assumed, or indeed not even thought about in explicit terms. In this context, Deb did not feel the need to differentiate himself according to homosexual or heterosexual identification.

Given the general lack of recognition of homosexuality among most of the people Deb encountered in his daily life, the idea that he should ascribe himself an explicit homosexual identity because he was in a sexual relationship with Bishwas did not make much sense. Being heterosexual, as it were, was a way for Deb to signify his commitment to the (implicitly hetero) normative. This did not preclude his involvement with CC or calling himself gay, but gay here did not perform as a narrow, neo-colonial form of identity. Indeed Leo's uncomprehending bemusement at Deb's account served as a stark illustration that 'gay' in this context was far from being unproblematically aligned to any sentiment that might be called 'Western gay identity.' The scenario described here is not an either/or situation – repressed or liberated – in or out – but rather a sexual subjectivity achieved in ambiguous relation to such dichotomies.

Reflecting on this experience, it is pertinent to stress that within the nominal West too same-sex sexuality may be far from universally and unproblematically circumscribed within obvious identity matrices. While idiosyncratic processes in nineteenth century Europe came to define gendered sexual object choice as *the* marker of sexual orientation, a range of other possibilities of sexual expression remained (and remain) de-emphasized as constituents of distinct sexual subjectivities (Sedgwick, 1991). Eve Kosofsky Sedgwick observes that:

> At the same time that this process of sexual specification or species formation was going on [...] less stable and identity-bound understandings of sexual choice also persisted and developed, often among the same people or interwoven into the same systems of thought
>
> (Sedgwick, 1991: 9).

Sedgwick emphasizes the contradictory effects of overlapping discourses and practices as they compel asymmetrical power relations in a homo/heterosexual dichotomy amidst the context of diverse and less defined expressions of sexuality. In this vein, the moral orthodoxies flowing from definitions of homo- and heterosexuality that ossified during the course of the nineteenth century both drew on and filtered out to colonial contexts, informing ethnographic accounts and Imperial legislature (Bleys, 1996). Such practices were intrinsically bound up with the racial and cultural essentialisms that demarcated colonized and colonizing peoples. In this context, sexuality came to stand as a symbol of European restraint and virtue while, as a necessary counterpoint, it also represented the degeneracy of colonial subjects (depicted as sexually voracious, immoral, etc.)

morality and same-sex sexuality in India: colonial legacies and contemporary practices

While conscious of the limitations of blaming all of contemporary India's social problems on a post-colonial legacy, Ruth Vanita and Saleem Kidwai have argued that 'a homophobia of virulent proportions came into being in India in the late nineteenth and early twentieth centuries and continues to flourish today' (Vanita and Kidwai, 2000). The period after 1857, when the British defeated the Indian 'rebellion,' resulted in the consolidation of Imperial rule in India (replacing the heretofore governance of the East India Company). The era saw the introduction of extensive British law and policy, including legislation designed to police the sexual lives and perceived immoralities of Indian subjects. This was an effective means to demarcate differences between the rulers and the ruled, inculcating a colonized mentality at the most intimate level (Nandy, 1983; Bhaskaran, 2002). This was typical of processes via which the erotic lives of colonized people were judged as morally wanting (Bleys, 1996; Weeks, 2003).

In India, same-sex sexuality was first criminalized in 1861 with the introduction of the British anti-sodomy law as section 377 of the Indian penal code. While this marked a liberalization of policy in Britain (replacing the mandatory death sentence with ten years' imprisonment), in India this represented a new form of legal discrimination. In practice the law has rarely been used (Bhaskaran, 2002), although prosecutions have taken place – most recently and notably with charges brought against men working in a community project for men who have sex with men in the Indian city of Lucknow (Cohen, 2005; Joseph, 2005). While prosecution of sodomy may be rare (in India and elsewhere), criminalization nevertheless establishes a moral climate of sorts, with concomitant effects for same-sex sexual practice.

Among more pervasive effects of section 377 the suppression of pre-existing practices of same-sex desire in India via the imposition of complex and ambiguous homophobic ideologies has arguably been most profound (Vanita and Kidwai, 2000). Scott Kugle (2002) has explored the post-1857 repression of homoerotic themes in north India Persian–Urdu poetry. A new morality saw the replacement of accounts of male-to-male love stories in favour of an 'invented tradition' of Islamic (heterosexual) manliness. This move was closely tied to the British displacement of Mughal imperial power in India, which narrowed the legitimate scope for the expression of same-sex love. The new literature served the ideological purposes of British imperialists, who sought to depict Mughal rulers as fanatics oppressing a passive Hindu majority. Images of Islamic manliness were intrinsic to the construction of a counter-discourse of Hindu effeminacy and incapacity. Thus in creating their own imperial ideology the British conceived themselves as liberators of the oppressed whilst still casting Hindus as intrinsically incapable (but nevertheless better off under the more benign rule of the new empire).

Complex perspectives on male homosexuality emerged in this context. Mrinalini Sinha's account of the development of ideas concerning the perceived effeminacy of Bengali men indicates variant perceptions of the relation between gender, sexuality and power:

> The British demonstrated the Bengali's lack of 'manly self-control' in arguments about the excessive sexual indulgence of the Bengali male, represented by the premature consummation of marriage as well as by the overly sexual atmosphere of the Bengali home that allegedly led to such practices as masturbation. As further indication of the unevenness in the intersection of metropolitan and colonial contexts, moreover, the elaboration of the debilitating sexuality of the Bengali male intersected in complex ways with the elaboration of a distinct homosexual personality in contemporary medical and scientific discourses in Britain. It is interesting, for example, that in the popular colonial imagination homosexual practices were associated with the favoured 'manly' and 'virile' native races rather than with the effeminate Bengali, the object of colonial derision. [...] The colonial focus on effeminacy to distinguish the sexually enervated Bengali male from the sexually virile frontier tribesmen in India exposed the contradictions of a discourse that attempted to link homosexual practices with a distinct homosexual personality defined in terms of effeminacy and lack of manly virility.
>
> (Sinha, 1997: 19)

Despite the contradictions highlighted by Sinha, this account can be read as replicating imperial ideologies regarding homosexuality. The 'manly homosexual,' while apparently incompatible with prevailing beliefs about gender and sexuality, bears the residual imprint of the kind of traditions Kugle describes as repressed by British ideology (albeit in the case Sinha describes this is not restricted to a Persian–Urdu tradition). In this instance, the fascination with the non-effeminate homosexual indicates the strength of prevailing (European) beliefs about gender and homosexuality. The 'manly' homo- subject was only of interest because he was seen as an anomaly. He was cause for comment because existing ideologies equating effeminacy and homosexuality were (and are) so strong, rather than because he instigated a re-evaluation of such beliefs *per se*. Similarly, the putative heterosexuality of the Bengali was equally troublesome because his construction as an effeminate colonial subject indicated homosexuality within prevailing ideologies. Despite a lack of apparent association with actual homosexuality, the Bengali was still cast as effeminate and the strength of this as a disparaging subjugation rested on the correspondence between male effeminacy and homosexuality. Thus, while the scenarios described did not neatly match hegemonic homophobic categories, the fact that each archetypal man was marked out as in some ways 'troublesome' indicates the prevailing strength of homophobic beliefs in colonial ideologies. The stigma was made to apply, despite empirical dissonance, because such a reading of gender and sexuality suited the imposition of British power.

The subjects conceived in British colonial homophobic discourses did not neatly match emerging European ideas of the distinct homosexual person because men who have sex with men in India were not differentiated from heterosexual practices in expected ways. This ambiguity translates into contemporary contexts; as Vanita observes, 'In India, most people have been, and many continue to be, married off at a very young age. Hence exclusive same-sex relationships are necessarily rare' (Vanita, 2002a: 3). Because of the great emphasis on marriage as a marker of adulthood in Indian culture, the consolidation of sexuality around ideas of individuated sexuality is less entrenched. Prejudice certainly informs discretion about same-sex sexual contact and relationships in India, as well as more explicitly homophobic brutalities. However, the technologies of power that produced the strongly individuated homosexual (and heterosexual) subject in nineteenth century Europe have not had the same effect in India. Simultaneous moral ambivalence about homosexuality informs a social context in which same-sex sexual subjectivity is far from uniformly common as an outcome of same-sex sexual relations.

practising ambiguity

One evening, an interviewee of mine, Abishek, had been travelling into Calcutta, after spending the day working a few miles out of town. The sky was heavy with imminent rain and he was daydreaming through the train window, his eye caught by the flickering speck of a red kite turning against the slate grey cloud in the distance. Soon, however, Abishek found his eye caught by something more interesting as he became aware of a good looking man who was staring at him from the corner of the carriage. Abishek's version of events unfolded thus:

> As more people came onto the train at the next station we were crowded further into the carriage. The man kept looking at me; soon he was next to me, pressed up against me. I could feel him rubbing up against me but wasn't sure. Was he touching me deliberately or was it just the crowd? At the next station I got down and the man did too. There had been an accident on the metro that day, and because of this I had to wait longer than usual at the station to catch my next train. While I was waiting the man approached me. We began to talk and then went for a walk outside the station. During the walk he made a sexual advance to me. It was interesting because when we discussed where we could have sex the man mentioned a number of *boudi's* houses where we could meet during the day. The idea was that the *boudi* would join in the sex.

Boudi is the Bengali word for an elder brother's or cousin's wife. It may be applied literally (to refer to a sister-in-law) or more generally, in 'fictive' terms, as a respectful word for a slightly older woman. The significance here is that the 'man on the train' was proposing a sexual experience involving Abishek, himself and a woman. Abishek was especially keen to tell me about this proposition because it was a stark illustration of something he knew I was interested in, the apparent accommodation of homosexual experience within a context that might look like heterosexuality. This complicates the premise of discrete hetero- and homosexual categorization. For Abishek the experience was, on one plane, explicitly homoerotic. However, he felt unsure about the other man's intentions, and by association his own, since despite the predominance of homosexual practices in his life he did not necessarily rule out attraction to women. This was especially so as he imagined that he might someday get married and as such did not foreclose the possibility of same-sex sexuality as intrinsic to his adult life.

This interstitial approach to homo- and heterosexual object choice has been well discussed in HIV/AIDS oriented work on male-to-male sexuality in non-Western countries (including India) (Khan, 1996). Such practices have been labelled by Peter Aggleton (1996) as 'behaviourally bisexual', although he acknowledges the contingency of such a framing: it potentially deflects attention from the contextual specificities where male-to-male sexuality may be taken as normative but un-emphasized in a subject's conceptualizations of self. Certainly within my own ethnography a number of men spoke in terms of attraction towards both men and women (perhaps privileging a different discourse in different contexts). However, I do not want to simply categorize such discourses as bisexual, as if such a label simply makes sense of these experiences. As Clare Hemmings (2002) has argued, bisexual subjectivity is characterized by performances within straight, gay or queer space (bisexual space being temporally and geographically limited). In these terms bisexuality is intrinsically partial, characterized by negotiations through, and passings within, hetero or homo contexts. Classifying the men I describe as bisexual would obscure their partial enactments of sexuality. The label would not only be meaningless to the men described, but would imply a coherent sphere of bisexual experience dissonant from the inherently fragmented and irregular sexual practices indicated.

I should stress that I do not want to develop a critical analysis of the potential utilities and problems of a bisexual classification in HIV/AIDS work. Rather, I want to relate the scenario I am describing to a broader conflation of homosexual and heterosexual discourses in Indian culture and society. This was noted by Abishek, who was keen to point out that his story was a particularly salient example because the subject of eroticism between a woman and her younger brother-in-law (to whom she is *boudi*) is a prevalent theme within Indian literature and culture. Gupta (2001), for example, has noted popular elaborations on this subject in songs, proverbs and jokes in Uttar Pradesh (North India). Rabindranath Tagore has written on a similar theme in respect of a specifically Bengali context, 'suggesting the eroticism inherent in the forbidden crossing of boundaries, where the woman often becomes attracted towards her brother-in-law' (Gupta, 2001: 202).

Within my own fieldwork, the connectedness between ostensibly distinct homosexual and heteronormative matrices is reiterated, as with Abishek's story, in the way a number of respondents spoke of erotic feelings towards archetypal *boudi* figures. These were sometimes actual sister-in-laws, but also commonly other women who were idealized or eroticized in such terms. A number of my respondents, for example (who were known to me as 'men who have sex with men' and explicitly 'sampled' in respect of such a self-identified attribute), spoke of how they were attracted towards women who were slightly older than themselves, for instance admiring the ample curves of their fuller bodies (as compared with the stereotypically thinner bodies of younger women).

Aside from attractions for *boudis*, more substantive sexual relations with women were seen as an important (and perhaps unavoidable) part of many respondents' lives. For some, family plans for their marriage were cause for concern (being seen as incompatible with a sexual orientation toward other men). For others, however, the prospect of marriage was eagerly anticipated, and my questions about potential tension between homo- and heterosexual intent in this regard were considered irrelevant or incomprehensible. Marriage, and the heterosexuality this implied, was seen as an inevitable and desired rite of passage into adult life. This did not preclude sex with men, and indeed many married respondents had male lovers – some feeling guilt and/or shame about this, others being less troubled.

The spatial arrangements of typical Indian family life intensify the scope for same-sex sexual contact. Khan has offered a reading of blurred sexual practice as relevant to both ambivalent sexual subjectivity and the context for sexual acts between men in India:

> An illustration of this is provided by an event that occurred whilst visiting a single-room home in New Delhi shared by the parents, four children, one a male in his mid teens, and an uncle of the children. The male teenager and the uncle were sharing a blanket while the female members of the household were getting on with the housekeeping. After a while it became clear what the two males were doing under the blanket, a behaviour totally ignored by the women. Subsequent discussion with the uncle revealed that during the night the two men would often masturbate together under their shared blanket, and on some occasions, the older would penetrate the younger when everyone was asleep. Because the behaviour was invisible, there did not seem to be any sense of shame or guilt. As the uncle put it, 'What can I do? I get body tension. He gets body tension. We are together. It just happens. We are friends.' When asked will he get married the same man replies 'Of course'. When asked would he have sex with women if he had the opportunity he replies 'Of course'. When asked has he had sex with other men, he replies 'Of course' under similar circumstances! (Khan, 1996: 171–172).

One of the interesting things in the above anecdote is the tension between Khan's description of the act as 'invisible' and his actually having witnessed it. The nature of the concealment was seemingly ambiguous. Thus, the women in the household went about their chores, *ignoring* what was going on beneath the blanket. In order to ignore something one has to perceive it (even if peripherally), which suggests that the kind of visualism Khan wants to convey is not so much what is or is not *seen* as what is or is not acknowledged. For Khan, this interpretation augments the lack of shame his subject feels, this reading extending through a constitution of India as a 'shame' rather than a 'guilt' culture (Khan, 1996). This perception is apparently founded on a popular reiteration of Indian 'persons' as constituted through socially oriented rather than individual understanding of self. Thus, social shame is more significant than individual guilt. This interpretation implies an indistinct context for male-to-male sexuality, in which such relations only become problematic when allied to claims for socially recognized sexual identities. Otherwise sexuality between men is known about and implicitly tolerated. This complex moral scenario derives from and supports simultaneously pejorative and permissive attitudes towards sex between men. Ostensibly contradictory perceptions of India as a society in which male-to-male sex is both heavily stigmatized and implicitly accepted reflect this cultural ambiguity.

alterity, morality and change: non-identified men who have sex with men in the context of HIV/AIDS prevention

This paper has explored variant possibilities for male-to-male sexualities in a climate of ambiguous moral censure – contemporary India. I have argued that while homophobia (as a post-colonial legacy) has profound effects on people's lives, parallel ambivalence about same-sex sexuality simultaneously allows social latitude for sexuality between men. This fosters a context in which male-to-male sexual connections may have a weak relation to same-sex sexual subjectivity. This raises questions about the place of male-to-male sex within ostensibly heteronormative contexts – indicating ways in which (same-sex) sexuality is not always yoked to discrete homo-subjectivities (as Sedgwick has observed above).

Judith Butler (2000) has accented the necessity of conceiving lives that are lived beyond (heteronormative) symbolic order (or social law). She argues that such 'unorthodox' lives should not just be seen as the shadowy outer margins that represent the foreclosure of heteronormative kin and cultural boundaries, but as lives lived *within* the norm. Butler utilizes the myth of Antigone to signify difference imminent within orthodox kinship structures. She critiques Hegel and Lacan's interpretation of Antigone's tragic death/suicide consequent to her love for her brother, as an example of the *reassertion* of law. Rather Butler argues that Antigone represents the possibility of other love in the context of ostensibly heterosexual kinship (Butler, 2000). While in the case of Antigone such sexuality can be seen as suppressed by the weight of conventional morality, it can also be made to stand for less defined sexual possibilities imminent within more apparently orthodox social connections. Male-to-male sexuality in India can be understood in these terms, as a sexual possibility intrinsically bound up with heteronormative contexts, rather than as necessarily separated out as an individual and social sense of self, identity or sexuality.

In putting forward such an idea I do not want to adhere to sociological vagueness. Men who have sex with men who enact sexual relationships through the conventions of heteronormative sociality are not necessarily actively challenging homophobia. Typically, such relationships are concealed out of awareness of the weight of homophobic values. Certainly, many men in Calcutta, for example, were all too conscious of the social stigma and shame they might suffer if they publicly identified themselves as men who have sex with men. Homophobia is real in its effects in India, and many men and women suffer explicit prejudice and violence as a result. With this in mind I do not want to propound a foreigner's travelogue convention of male-to-male sexuality in India as simply free and tolerated. Nevertheless, the social construction of same-sex sexuality in India is such that the effects of homophobia are uneven, which allows considerable social latitude for un-emphasized and un-scrutinized same-sex sexual connections. This results in ambiguities in the way that many men and women conceive their sexuality, since the explicit moral authorities that have consolidated the same-sex sexual subject as a social category in nominally Western contexts have not commonly engendered the same sexual subjectivities in Indian society.

This complex scenario is especially relevant to HIV/AIDS work in the subcontinent. Unlike most Western countries, HIV/AIDS in India (as in many non-Western contexts) has been primarily associated with heterosexual sex. Thus despite homophobia, men who have sex with men were not commonly marked out as a stigmatized HIV/AIDS risk group as the epidemic took hold in the subcontinent in the mid 1990s. Rather, in this context as in others, activists had to fight for recognition of such men as vulnerable to infection (Parker *et al.*, 1998; Joseph, 2005). This suggests a very different kind of moral context to Western countries. Although one could argue that the strength of homophobia in India is such that the existence of homosexual men is suppressed and denied, the low social profile of men who have sex with men also arguably emerges from a context of moral ambivalence, in which homophobia has not had sufficient cultural depth and scope to produce a broadly recognized constituency of men who conceive themselves as same-sex sexual subjects – with resulting lack of social recognition of male-to-male sexuality (either positive or negative).

This raises critical problems for the promotion of health. Denial that men who have sex with men exist in India has often led to reductive underestimates of their presence and poor understanding of their epidemiological significance and vulnerability in HIV/AIDS work (Khan, 1996). Activism premised on gay identity has been important in raising awareness that men who have sex with men are present in India (Vanita, 2002a). Alternatively other activists have asserted the need to recognize putatively indigenous identity categories – since gay identity is arguably not relevant to the majority of men who have sex with men (Khan, 2001). While this work has been important in diversifying the breadth of HIV-related intervention, it tends to replicate a perception of sexual subjects as only knowable through classification. This is understandable, as advocacy strategies inevitably have to assert the empirical existence of men who have sex with men – and categorizing people is one way of doing this. However, such strategies limit the agenda of HIV/AIDS work, narrowing capacity to address the actual circumstances of the many men in India for whom categories of same-sex sexuality are not salient.

A more subtle position is being advocated in Indian legal contexts. Petitions in India for same-sex sexual rights have focused on the repeal of section 377. The argument for repeal has been made by activists in the name of privacy, rather than sexuality *per se.* This is reflected in a parallel strategy, aimed at claiming rights to privacy under Article 21 of the Indian constitution, which includes the right to self-determination of identity as inviolable, integral and personally intimate, rather than publicly identified. Katyal has offered a cogent analysis of the significance of this strategy:

> This interpretation of personal identity mirrors a similar preoccupation in gay rights circles in India with the right to determine both the interior and exterior aspects of a person's sexuality [...] In other words, a strategy that focuses

> primarily on the right to privacy *before* identity-based protections allows an individual the autonomy to determine his or her sexual identity and preferences, instead of forcing them to adopt a particular identity to access constitutional protection. In other words, by demanding the right to privacy, Indian activists are actually demanding the right to deliberate- and determine- the interior and exterior aspects of their sexual identity, for themselves, and not as a prerequisite for a particular legal entitlement.
> (Katyal, 2002: 166)

The petition described arises from a realization that in India identification on grounds of same-sexual orientation is inimical to many people. What is offered here is a way of conceiving the rights and needs of a broad population of same-sex sexual subjects, while at the same time comprehending that many men will not find such categorization comprehensible or meaningful. This strategy reiterates the pervasive social presence of male-to-male and female-to-female sexualities without falling back on superficial understandings of stigmatized groups. HIV/AIDS policy and programming can draw on this kind of approach. This would offer a basis from which to conceive the silences, repression and social latitude that characterize the experience of same-sex sexuality in India – these being the most pressing issues compounding vulnerability to HIV infection, as they both allow considerable scope for sexual activity while curtailing the reach of health promotion messages aimed at identifiable men who have sex with men. Taking account of ambiguous moral censure regarding same-sex sexuality in India enables a better understanding of the irregular effects of homophobia. This in turn enables more subtle analysis of, and programmatic response to, social processes that compound the socially indistinct presence of men who have sex with men, who face very real prejudices and risks.

acknowledgements

I thank Deep Purkayastha, Clare Hemmings, Henrietta Moore, Pawan Dhall, Peter Aggleton and the un-named research participants for their valuable contributions towards this paper.

author biography

Paul Boyce received a Ph.D. from the London School of Economics and is currently an ESRC postdoctoral research fellow at the Thomas Coram Research Unit, Institute of Education, University of London.

references

Aggleton, P. (1996) 'Introduction' in Aggleton, P. (1996) editors, *Bisexualities and AIDS: International Perspectives*, London: Taylor & Francis.

Bhaskaran, S. (2002) 'The politics of penetration: section 377 of the Indian penal code' in Vanita, R. (2002) editor, *Queering India: Same-Sex Love and Eroticism in Indian Culture and Society*, New York: Routledge.

Bleys, R.C. (1996) *The Geography of Perversion: Male-to-Male Sexual Behaviour Outside the West and the Ethnographic Imagination, 1750–1918*, London: Cassell.

Boyce, P. (forthcoming) *Conceiving Kothis: Role, Agency and Anthropology in the Construction of Male-to-Male Sexualities in India* – medical Anthropology.

Butler, J. (2000) *Antigone's Claim: Kinship Between Life and Death*, New York: Columbia University Press.

Cohen, L. (1995) 'The pleasures of castration: the postoperative status of Hijras, Jankhas, and academics' in Abramson, P.R. and Pinkerton, S.D. (1995) editors, *Sexual Nature/Sexual Culture,*, Chicago: The University of Chicago Press.

Cohen, L. (2002) 'What Mrs. Beshara saw: reflections on the Gay Goonda' in Vanita, R. (2002) editor, *Queering India: Same-Sex Love and Eroticism in Indian Culture and Society*, New York: Routledge.

Cohen, L. (2005) 'The Kothi wars: AIDS cosmopolitanism and the morality of classification' in Adams, V. and Leigh Pigg, S. (2005) editors, *Sex in Development: Science, Sexuality, and Morality in Global Perspective*, Durham: Duke University Press.

Donham, D. (2005) 'Freeing South Africa: the "Modernization" of male-male sexuality in Soweto' in Robertson, J. (2005) editor, *Same-Sex Cultures and Sexualities: An Anthropological Reader.*, Malden: Blackwell.

Gupta, C. (2001) *Sexuality, Obscenity, Community: Women, Muslims and the Hindu Public in Colonial India*, New Delhi: Permanent Black.

Hemmings, C. (2002) *Bisexual Spaces: A Geography of Sexuality and Gender*, New York: Routledge.

Joseph, S. (2005) *Social Work Practice and Men Who Have Sex With Men*, New Delhi: Sage.

Kakar, S. (1997) *Culture and Psyche*, New Delhi: Oxford University Press.

Kakar, S. (2002) 'Clinical work and cultural imagination' in Vaidyanathan, T.G. and Kripal, J.J. (2002) editors, *Vishnu on Freud's Desk: A Reader in Psychoanalysis and Hinduism*, New Delhi: Oxford University Press.

Kartrak, K.H. (1992) 'Indian nationalism, Gandhian "Satyagraha" and representations of female sexuality' in Parker, A., Russo, M., Sommer, D. and Yaeger, P. (1992) editors, *Nationalisms & Sexualities.*, New York: Routledge.

Katyal, S. (2002) 'Exporting identity' *The Yale Journal of Law and Feminism*, Vol. 14, No. 1: 97–175.

Khan, S. (1996) 'Under the blanket: bisexualities and AIDS in India' in Aggleton, P. (1996) editor. *Bisexualities and AIDS: International Perspectives*, London: Taylor & Francis.

Khan, S. (2001) 'Culture, sexualities and identities: men who have sex with men in India' in Sullivan, G. and Jackson, P.A. (2001) editors. *Gay and Lesbian Asia: Culture, Identity, Community*, New York: Harrington Park Press.

Khanna, A. (2005) 'Beyond sexuality (?)' in Narrain, A. and Bhan, G. (2005) editors, *Because I Have a Voice: Queer Politics in India*, New Delhi: Yoda Press.

Kugle, S. (2002) 'Sultan Mahmud's makeover: colonial Homophobia and the Persian–Urdu literary tradition' in Vanita, R. (2002) editor. *Queering India: Same-Sex Love and Eroticism in Indian Culture and Society*, New York: Routledge.

Nandy, A. (1983) *The Intimate Enemy: Loss and recovery of self under colonialism*, New Delhi: Oxford University Press.

Osella, C., Osella, F. and Chopra, R. (2004) 'Introduction: towards a more nuanced approach to masculinity, towards a richer understanding of South Asian men' in Chopra, R., Osella, C. and Osella, F. (2004) editors, *South Asian Masculinities: Context of Change, Sites of Continuity*, New Delhi: Kali for Women.

Parker, R., Khan, S. and Aggleton, P. (1998) 'Conspicuous by their absence' Men who have sex with men (msm) in developing countries: implications for HIV prevention' *Critical Public Health, HIV/AIDS Symposium Issue*, Vol. 8, No. 4: 329–346.

Reddy, G. (2004) 'Crossing "Lines" of subjectivity: the negotiation of sexual identity in Hyderabad India' in Srivastava, S. (2004) editor, *Sexual Sites and Seminal Attitudes: Sexualities, Masculinities and Culture in South Asia*, New Delhi: Sage Publications.

Row Kavi, A. (1999) 'The contract of silence' in Merchant, H. (1999) editor, *Yaraana: Gay writing from India*, New Delhi: Penguin Books.

Rubin, G. (1999) 'Thinking sex: notes for a radical theory of the politics of sexuality' in Parker, R., and Aggleton, P. (1999) editors, *Culture, Society and Sexuality: A Reader*, London: UCL Press.

Sedgwick, E.K. (1991) *Epistemology of the Closet*, Harmondsworth: Penguin Books.

Sinha, M. (1997) *Colonial Masculinity: The 'manly Englishman' and the 'effeminate Bengali' in the Late Nineteenth Century*, New Delhi: Kali for Women.

Srivastava, S. (2004) 'Introduction: semen, history, desire and theory' in Srivastava, S. (2004) editor, *Sexual Sites, Seminal Attitudes: Sexualities, Masculinities and Culture in South Asia*, New Delhi: Sage Publications.

Strong, T. (2002) 'Kinship between Judith Butler and Anthropology? A review essay' *Ethnos*, Vol. 67, No. 3: 401–418.

Uberoi, P. (1996) 'Problematising social reform, engaging sexuality, interrogating the state' in Uberoi, P. (1996) editor, *Social Reform, Sexuality and the State*, New Delhi: Sage Publications.

Vanita, R. (2002a) 'Introduction' in Vanita, R. (2002a) editor, *Queering India: Same-Sex Love and Eroticism in Indian Culture and Society*, New York: Routledge.

Vanita, R. (2002b) 'Homophobic fiction/homoerotic advertising: the pleasures and perils of twentieth-century Indianness' in Vanita, R. (2002b) editor. *Queering India: Same-Sex Love and Eroticism in Indian Culture and Society*, New York: Routledge.

Vanita, R. and Kidwai, S. (2000) editors, *Same-Sex Love in India: Readings from Literature and History*, New York: Palgrave.

Warwick, I., Douglas, N., Aggleton, P. and Boyce, P. 'Young gay men and HIV/AIDS. Towards a contextual understanding of sexual risk' *Sex Education*, (2003) Vol. 3: 215–229.

Weeks, J. (2003) *Sexuality: Second Edition*, London: Routledge.

doi:10.1057/palgrave.fr.9400282

83 revolutionary marriage: on the politics of sexual stories in Naxalbari

Srila Roy

abstract

Marriage practices, the dynamics of interpersonal relationships and the politics of sexuality are relatively under-researched themes in the study of Bengali communism. Historical scholarship on the revolutionary politics of the extreme left Naxalbari *andolan* of the late 1960s–1970s, the object of this piece of study, is no exception. The article engages with women and men's narratives on the practice of 'revolutionary' marriage in the movement through the prism of contemporary popular memory studies and narrative analysis. Drawing on field interviews with middle-class male and female activists, the article draws attention to the contestatory nature of marriage in the collective memory of the movement. Narrative contestations over marriage in the Naxalite movement underscore, I argue, a tension between a utopian ideal of transgressive interpersonal relations and dominant middle-class codes of sexual morality. At the same time, individual attempts to 'compose' (in storytelling) socially recognizable and acceptable subject positions are grounded upon the silencing and abjection of more risky memories. Given the discrepancies and contradictions within the narrative repertoire from which individuals construct their identities, these 'marriage stories' are a tremendous resource for investigating the politics of love, sexuality and subject-formation in middle-class Bengali society.

keywords

sexuality; identity; narrative; memory; revolutionary movements

(99–118) © 2006 Feminist Review. 0141-7789/06 $30 www.feminist-review.com

revolutionary marriage: on the politics of sexual stories in Naxalbari[1]

When in love
Do not become a flower
If you can,
Come as the thunder.
I'll lift its roar to my breast
And send forth the battle cry to every corner.[2]

These lines are part of a poem written by young Bengali Naxalite, Murari Mukhopadhyay. In young middle-class men and women's dreams and desires for a utopian world, the idiom of love found new usage. The imagery of love fused new aspirations and passions for revolutionary violence, subverting its erstwhile 'bourgeois' underpinnings. Love was (re)imagined as a form of comradeship that transpired only in the revolutionary cause. What is more significant is that *bhalobasha* (love) and desire for 'revolution' formed the basis for the oneness of the revolutionary couple; their love and unity is facilitated by *biplab* or 'revolution'. The rewriting of the institution of 'marriage' in the Naxalbari *andolan* was similarly premised on an equality of partnership, forged through an isomorphism of *biplab* with *bhalobasha*. Revolution entailed, it seems, a radically altered organization of romance and relationships that had significant implications for the problematic of gender.

The Naxalbari *andolan* began as a peasant uprising in northern West Bengal in 1967, led by a dissident group of the Communist Party of India (Marxist). Members of this group and their sympathizers came to be known as Naxalites. The Naxalites declared a 'people's war' against the Indian state structured on the Maoist model of protracted armed struggle. Armed with a copy of Mao's *Little Red Book*, middle-class students, who formed a wide base for the movement, left the city in order to 'integrate' with the peasantry and become 'de-classed'.[3] The political line of *khatam* or the individual annihilation of 'class enemies', first instigated against landowners in rural areas, escalated into what has often been referred to as an orgy of violence. Small guerrilla units primarily of men indiscriminately killed anyone from traffic policemen to local schoolteachers as representatives of the state. The movement was finally crushed in 1971 under severe state repression and partly due to the political misgivings of the Party. Stories of young idealist men being brutally tortured and shot by the police have been the most sustained component of the Naxalbari legacy.

Although contemporary Naxalite and Maoist groups operate in other parts of India, Bengal has never seen a resurgence of Naxalite violence after the events of the 1960s–1970s. Yet this movement forms an intricate thread in the lived memory of the city of Calcutta, and continues to be one of its dominant legends.

1 A shorter version of this article was presented at the Postgraduate South Asia Seminar Series, Department of Anthropology, London School of Economics and Political Science 2004

2 From Banerjee (1987)

3 The idea of 'declassing' the self in order to revoke the ideological distance between the 'intellectual' and the masses has a long-standing tradition in middle-class Bengali Marxist politics. Becoming 'de-classed' meant, for the *bhadralok* Marxist, the sacrifice of customary material privileges and aspirations, beginning with the abandonment of domestic life and responsibilities. See Dasgupta, 2003.

This article explores narratives of 'marriage'[4] and the politics of interpersonal relations and sexuality in Naxalbari, a problematic that has found little or no attention in the movement's historiography (Banerjee, 1984; Ray, 1988). In tracing the contestatory nature and shifting meanings of 'marriage' in collective memory, I am particularly concerned with the narrative and discursive repertoires from which stories are told, and the forms of identity they 'compose'. The telling of 'marriage-stories' can thus be understood as one among a range of 'narrative practices' (Redman, 1999) in and through which a particular version of the subject is mobilized in response to existing power relations and forms of social recognition. I concentrate on the narratives of middle-class men and women activists.[5]

My approach to narrative is less as a text than as 'social actions' productive of meanings and images of community and self. To this extent, I draw on Ken Plummer's (1995) idea of a 'sociology of stories', an approach that seeks to go beyond the structural or textual work that narrative performs to the social, cultural and political role of stories in everyday life. I am equally interested in the discursive repertoires from which stories are generated, the intersubjective domains in which they are told, and the historical and sociological work they perform. A concomitant area of interest is that of subjectivity.[6] Here I focus on a very specific problematic, that of 'composure' or the narrative function of producing a coherent subject-position that is recognizable and 'livable' (Dawson, 1994). The question of recognition (and its intimate relationship with subjectivity) goes back to the idea of the 'cultural circuit' (Johnson, 1982) where narratives are understood as located within a feedback loop between personal and public stories (Summerfield, 2000).

I begin by briefly historicizing the practice of marriage in the Marxist tradition, drawing attention to the anxieties around female sexuality in radical left politics. The next section lays down the discursive framework of 'revolutionary marriage' in Naxalbari, and the inherent instabilities of this discourse. I then move to a discussion of two narratives that draw on dominant discursive renderings of 'revolutionary marriage' to compose self-identities that are potentially fragile.

politicizing marriage

Free choice marriages and consensual unions have been a common characteristic of most communist movements. This tradition can be traced back to the early Marxist analysis of the 'women's question' that included a critique of monogamous marriage, the basis of the bourgeois family. Socialism, it was believed, would lead to a higher form of monogamy, one that was not marred by property relations.[7] 'Socialist monogamy' (Evans, 1992) was the ideological motivation behind the progressive redefinition of conjugality in communist China.

4 I put marriage in quotation marks given that the manner in which it was practiced by the Naxalites (or at least by a large number of them) does not comply with the conventional practice of marriage (civil or religious). I also wish to draw attention to the underlying politics of this act of naming; most people chose to use the term marriage ('beye') to characterize relationships of the time even though they existed outside the bounds of marriage. This is dealt with in greater detail further below.

5 These narratives were obtained through interviewing. The study relied on a 'narrative approach' to interviewing informed by the principles and ethics of feminist interviewing practices (see Gluck and Patai, 1991; Summerfield, 1998; Hollway and Jefferson, 2000; Lawler, 2002). The narrative approach crucially differs from the traditional question-answer interview format in that it rejects standardized questions for open ended, non-directive ones in order to encourage the telling of stories rather than to obtain a factual report.

6 The narrative constitution of the self has emerged as a legitimate area of scholarship on its own (see, for instance, Ricoeur 1991; Somers 1994). Feminists have been especially attracted to the possibilities

The 1950 Marriage Law defined marriage as one based on mutual love, free will and consent, free from concerns of property and wealth. At the same time, as demonstrated by Evans, this socialist ideal of monogamy transformed into a regulative one that reinforced traditional norms of sexual morality, the custodians of which were invariably women.

In revolutionary Marxist and Maoist groups closer to home (including the Naxalites), marriage and sexuality have been dealt with in less structured ways but have displayed similar disciplinary mechanisms within a rhetoric of revolutionary change. Women's narratives of the Telangana People's Struggle, a major communist-led peasant insurrection, show how women continued to be the bearers of tradition, and were consequently the objects of social policing (Kannabiran and Lalitha, 1989). A Marxist economic determinism relegated questions of marriage, sexuality and family to the private domain while re-establishing culturally prescriptive power differentials between men and women. The experience of the Srikakulam movement, a contemporary of Naxalbari, displays similar anxieties to questions of interpersonal relationships and sexuality. The leadership of the movement was often caught between the ideal of comradeship and pre-existing patriarchal norms that extended disciplinary control over individual lives (Vindhya, 1990; see also Vindhya, 2000 on the sexual politics of contemporary radical groups). Vindhya's observations that the Party had no coherent policy towards the organization of interpersonal relations and that the socialist ideal of gender equality had not been conceived of in clear terms, rings true for the Naxalites. Like her, several of my interviewees explained this conceptual lack in terms of the temporal structure of the movement, saying that it was too short-lived for the development of any conscious policy. Decisions with regard to marriage and divorce were made, in the case of both Srikakulam and Naxalbari, by local area committees, some of which (in the case of Naxalbari) consisted of nothing more than three members, usually all male.

In the Bengali Marxist tradition, members of the undivided Communist Party of India (the CPI) are generally credited with a degree of radicalism in experimenting with social relationships, although the private is silenced in most memoirs of the period (see Sen, 2001b; Lahiri, 2001). Several women who joined the Party between the late 1930s and early 1940s were from wealthy, conservative families, and lived in a commune run by the political activist Manikuntala Sen. This commune hosted both male and female activists and is often cited as evidence of the Party's potential to break gender barriers (Ray, 1999, n186). Neither the CPI nor the CPM (the Communist Party of India, Marxist) would entertain such a possibility today (*ibid*). Members of the women's wing of the ruling CPM[8] have a contradictory relationship with the Party. The Party often occupies the position of familial authority in arranging the marriages of women activists, and even presiding over issues of divorce. At the same time, women cannot politicize power structures within the family that are still considered

of a narrative approach for an exploration of gendered subjectivity. See Passerini *et al.*, 1996; Summerfield, 1998, 2000; Lawler, 1999; Redman, 1999; Byrne, 2003.

7 In his *Origins of the Family, Private Property and the State*, Engels (1884) spelt out the economic foundations of monogamous marriage, and its relationship to the production of capital and private property. Bourgeois monogamy necessitates, Engels argued, the production of heirs by women in order to preserve wealth and property in the hands of individual men. The key to women's liberation was the termination of the bourgeois family together with participation in social production. This classical Marxist approach to the 'women's question' held a number of relevant insights for Communist women like Alexandra Kollontai who argued beyond economic considerations to issues of sexual morality and emotions in discussions of women's emancipation. However, in general classical Marxism remained limited by its materialist basis of analysis. The other major trend that attracted feminists during the 1830s and 1840s was that of the utopian socialists who envisaged a new world order. Utopian movements have historically been more open to sexual experimentation than revolutionary movements have (Poldervaart, 2000). Like Engels' 'scientific theory' of women's subordination, utopian socialists located women's subordination in the sexually oppressive bourgeois patriarchal family, with its origins in the development of private property. Utopian socialists like the 19th-century British Owenites encouraged experimentation in changing the forms of family, childcare, and marriage. However, their efforts to redefine sexual and conjugal practices remained limited by their largely male-defined sphere of vision. The question of women's emancipation eventually became marginalized within the contest between gender and class (Taylor, 1983). Elsewhere in America and Germany, early 20th-century socialist-feminists faced political isolation and even hostility for their views on sexuality. Even in the 1970s student movements, a commitment to Marxist politics foreclosed discussions on family alternatives and sexuality (Poldervaart, 2000).

8 The Communist Party of India (Marxist) (CPM) was formed in 1964 after a split in the Communist Party of India. In West Bengal, the CPM-led Left Front government came to power after the 1977 emergency (partly owing to its success in crushing the Naxalites) and has never lost since.

'personal matters' within the organization (Ray, 1999: 81). In the case of both organizational and radical politics, the personal is negotiated in contradictory ways – relegated, at times, outside the domain of the political while constituting, at other times, the very object of a disciplinary gaze.

The *bhadralok* Naxalites,[9] one could say, inherited their radicalism from their Bengali communist predecessors and their Chinese counterparts in that they rejected the institution of marriage, especially that of arranged marriages, for companionate marriages based on love and mutual respect. Yet narratives on 'marriage' in Naxalbari reveal similar contradictions and anxieties about sexuality and gender that seem endogenous to radical left politics.

discourses on marriage: dominant and contradictory

'Marriage' remains a sticky issue in the history of the movement. For some, the organization of marital relations is evidence of the movement's empowering and progressive potential. For a small minority, by contrast, it demonstrates the Party's fundamental conservatism with regard to issues of sexuality and gender. In the narratives that I consider below, we can see the workings of at least three discourses on marriage – a dominant discourse of 'revolutionary marriage' and two contradictory renderings of it. While there is no Party documentation on the practice of marriage, an 'official' version can be identified in normative constructions of revolutionary marriage in male/female narratives and in literature as well. In this discourse, movement participants defied the institution of marriage as they did with all other social institutions, displaying their revolutionary zeal and progressive nature. This is a discourse of radical change and dramatic rupture in the manner in which 'revolutionary' marriage is constructed *vis-à-vis* arranged marriage, the dominant form of marriage in middle-class Bengali society. The rejection of traditional arranged marriages to emphasize partnerships based on love, equality and comradeship is accompanied by a rejection of both the religious and civil nature of marriage. In her short herstory of Naxalbari, Kalpana Sen (2001a), an ex-activist, argues that the Naxalite (re)definition of interpersonal relationships signaled not only a transgression of societal norms but the breaking of gender barriers for women.

Within the literary imagination of Naxalbari, love and romantic relations are infused with similar meanings. Much of this literature including Mahasweta Devi's (1997) acclaimed *Mother of 1084* emphasizes comradeship between partners, a chief component of true revolutionary love of the time. The relationships between activists Brati and Nandini in *Mother of 1084*, and between Bibi and Antu in Bani Basu's *Antarghat* (The Enemy Within, 2002) are characterized by friendship and mutuality that give meaning to a form of love-as-comradeship.[10] However, an egalitarian rhetoric of 'comrade' does not mitigate the literary representation of an idealized heroic masculinity and a dependent femininity that draws meaning from the romantic (male) other. Saibal Mitra's *Manabputri* (1993), a fictionalized account of Mary Tyler,[11] represents a form of love that is closer to devotional love than to a form of comradeship. The female protagonist, Katherine, a young English woman who becomes embroiled in radical left politics through her love-interest Dipayon, embodies middle-class virtues of husband devotion and shame (*lajja*).[12] Romance in revolution, expressed as a form of devotion or as friendship, is often governed by normative gender identities and hegemonic cultural codes.

How, then, was 'revolutionary marriage' practiced in the movement? Individuals could declare themselves husband and wife through an exchange of Mao's Red Book in front of the Party. More commonly though, individuals simply informed the Party that they were married in order to be recognized as such. The rejection of societal norms was not, however, absolute. Some did have a registered marriage. Ironically enough, in Mitra's novel, Katherine and her Naxalite partner are married in Kolkata's Kalighat temple!

Although it appears that norms and customs were often broken or maintained in accordance with individual choice, the Party, I would argue, played a significant role in the organization of interpersonal relations. Individual narratives are contradictory on this question, some emphasizing how the Party remained aloof from private lives that were secondary to the 'revolution'; others suggesting how Party discourses regulated everyday lives and practices in the movement. In many ways the Party became, I believe, the social self-conscious of the collective, substituting for the morality and legality of middle-class society in the 'underground'. The practice of the 'red book marriage' can be perceived as a substitution of one form of social institution (marriage as a legal contract) for another. At the same time, it is important to remember that the priority afforded to class struggle meant that everything else, especially love and sexuality (considered 'bourgeois' concerns), were regarded as secondary to this essential revolutionary task. As the lines of the poem above suggest, *bhalobasha* could *only* be validated for the sake of *biplab*. This also meant that the wider implications of 'marriage' such as the specific vulnerability of women as targets of sexual violence and domestic abuse were easily lost sight of.

The discourse of 'revolutionary marriage' has to contend with two other potentially contradictory discourses. As easily as these marital alliances were formed, so did they break. The short history of the Naxalbari movement speaks of a longer history of broken and betrayed relationships. Thus the first contradictory discourse is built around the fact that many relationships formed at the time did not last, a fact that renders a collective social experiment a large-scale failure. In emphasizing the movement's potential to break societal norms, men and women's stories struggle to understand and account for the unpleasant fact that

9 The term 'bhadralok' denotes a polite, civilized and respectable person, literally a 'gentleman'. Respectability for the *bhadralok* means a disdain for manual labour (the preserve of the *chotolok* or the lower classes), a privileging of education, access to a modicum of wealth, and maintaining a genteel and cultured lifestyle (Ray 1988, 2000). The *bhadralok* is not, however, a homogenous group of people, and there are marked distinctions within the group. The Naxalites mainly came from the ranks of the lower middle-class (most of who were refugees from the Bengal Partition) rather than those from *sahebi* backgrounds who are anglicized and upwardly mobile. The movement, as a whole, had a strongly entrenched lower middle-class character, dominated by a vernacular intelligentsia that was antagonistic towards the English speaking elite (Ray, 1988). The majority of my interviewees can be identified as being lower middle-class; some live in fairly impoverished conditions partly ascribable to political choices to do with lifestyle and career.

10 Devi's *Hajar Churashir Ma* (Mother of 1084 (1997)) is one of the more representative works on Naxalbari, and has been translated into several languages. Bani Basu's *Antarghat,* also centered on the movement,has been recently translated into English as *The Enemy Within*, 2002.

11 A 26-year old British woman who was arrested in relation to the famous Jaduguda Naxalite Conspiracy case, Mary Tyler was by far the media's favourite Naxalite (and the only woman to receive such enormous publicity). She spent five years in prison, which forms the basis of her memoir *My years in an Indian prison* (1997). In his interview, Saibal Mitra, a well-known ex-Naxalite and a fairly established contemporary writer, revealed to me how the character of Katherine and the novel at large is a fictionalized account of Mary Tyler.

12 In Bengali society, the *bhadramahila* or the gentlewoman is defined through middle-class codes of respectability, honour (*izzat*), passivity and sexual modesty, expressed in the acquisition of shame. *Lajja* – 'shame and modesty, attributes closely connected with virtue and respectability' (Ray, 2000:24) is a central signifier of middle-class femininity.

many of these relationships deteriorated (often in ugly ways), and the moral implications of such separations for *bhadralok* revolutionaries.

The second discourse that the dominant one on marriage has to contend with is far more threatening to middle-class codes of sexual respectability. The possibility of rejecting traditional norms that governed conjugal relations also meant the possibility of relationships existing *outside* the bounds of conjugality. In subtle and not so subtle ways, women and men's discussions on marriage try to negotiate this possibility of pre-marital sexuality and sexual licentiousness, which was encoded in the liberal reconstruction of 'marriage'. More than anywhere else, it is here that we see how these narratives of dramatic change and rupture are far more continuous with middle-class discourses of sexual morality.

Much of the discomfort that ex-activists feel in discussing 'marriage' is attributable to these contestatory discourses that effectively place the entire polemic of revolutionary marriage into question. In telling (and silencing) stories of relationships in the movement, narrators often strive to present a version of events that negotiates the contradictions and limitations within a dominant discourse of revolutionary marriage. In the narrative discussions that follow, we see how the subject draws on contending discourses to present a coherent version of events, and a version of the self that can elicit social recognition. Here I am drawing on Dawson's notion of 'subjective composure' which suggests that the narrator constructs a story from multiple possibilities that best allows her a sense of 'psychic comfort' through 'a subjective orientation of the self within the social relations of its world' (1994:22–23). Thomson (1994) similarly uses a concept of memory as composure that ties together cultural resources and individual psychic responses. The practice of 'composing' memories then has both a cultural and a psychological dimension to it. On the one hand, it is a thoroughly public one implicated within larger, more dominant versions of the past. On the other hand, remembering is vital to the work of identity, constituting the basis for the 'composure' of personal identities. Through the practice of composure, Thomson argues, individuals seek to transform risky, even traumatic memory into 'a safer less painful sense' (1994:10). Composure is, moreover, achieved through complex practices of repression and exclusion that nevertheless threaten its foundation:

> Composure 'based as it is on repression, and exclusion, is never achieved, constantly threatened, undermined [and] disrupted' (*ibid*).

Although Thomson does not entirely develop his observation, the point is significant – personal remembrance is based upon patterns of expulsion, rejection and repression that safeguard the coherency of the subject. At the same time, what is excluded threatens this very coherency; it threatens to destabilize our sense of self, and to produce *not* composure but what Summerfield (2000) calls *dis*composure. The threat of discomposure points to the inherent fragility of the constituted subject and to the unstable basis of its composure.

In the narratives that follow, we see how individuals struggle with a discourse of revolutionary marriage and its contradictory meanings to 'compose' a coherent subject position. Yet given the potential instabilities and contradictions that lie at the heart of these memories, the achievement of composure is all the more tenuous.

marriage as an abject zone

Kalyani[13] is one of the more prominent women Naxalites mostly because of her present-day political activism. I was introduced to her as to several other women through a friend's mother, an ex-activist herself and one who is highly respected amongst this circuit of women. In our very first conversation over the phone, Kalyani made her displeasure in talking about this period in her life explicit. She mentioned her reserve in discussing aspects of her past that she had not even revealed to her daughter. In fact, the interview began with Kalyani requesting anonymity, 'otherwise I won't say a lot. Anyway there are things I won't say'. Kalyani's narrative is significant for the liminal position that it occupies in the cultural memory of 'marriage' in the movement. As with women's narratives on marriage in general (and the male narrative that follows), it bears testimony to the contradictions that underlie a utopian vision of 'revolutionary marriage'; contradictions that, in turn, make the practice of remembering more risky today.

13 Name changed on request of anonymity.

Kalyani's short discussion confirms the dominant discourse of 'revolutionary marriage' insofar as it emphasizes the radically altered way in which marriage came to be understood and practiced in the *andolan*. She cites the case of an activist couple who, unlike some (but not all) Naxalites chose not to register their marriage even after the decline of the movement:

> Kalpana and Gautam have 'lived together'; they didn't get married. My first relationship, there was no registration or red book exchange. 'We just started' ... no problems. [...] You would just have to inform the Party as a so-called married couple. We challenged the institution of marriage, like we challenged all other institutions. [*What about the red book marriage?*] But how is that 'binding'? There is nothing binding in it. Today if I don't want to stay-lots of men married peasant women and then left them and came away. What do you understand by marriage? When an institution 'controls' you in some way. Here there is no control.[14]

There are two distinct senses of marriage that are at work in Kalyani's oral text. The first refers to the dominant way in which marriage is practiced in bourgeois *bhadralok* society – civil marriages that are legally registered and socially recognized. A second sense of marriage refers to the way in which it was practiced in the movement and which moreover 'challenged' the nature of marriage as a legal and social institution. Kalyani equates this second sense of

14 Interviews were conducted in English and Bengali. All translations from the original Bengali are mine. Words in single quotations were originally said in English. I interviewed Kalyani once which lasted for

roughly 4 hours. The interview with the couple that I go on to discuss spanned over four meetings, each interview lasting 3–4 hours.

marriage with 'living together' and disassociates it from any form of institutionalization. She emphasizes that 'Kalpana and Gautam have 'lived together'; they didn't get married'. Similarly, she speaks of her own relationship within the movement as one where there was no 'registration' or 'red book exchange'. This is revealing since she locates her relationship outside the contractual nature of the state/civil society *and* the Party.

While almost all male and female activists use the term 'marriage' in the context of the movement (irrespective of whether such marriages were legally recognized or not), Kalyani is one of the few who spells out a distinction between marriage and 'living together'. Given the uneasiness with which the concept of 'living together' sits on the Bengali middle-class (see, for instance, Basu, 2001), interpersonal relationships formed at that time are defined as conjugal ones even though individuals were very often living together. As the narrative below makes clear, even within the revolutionary community, 'living together' is associated with sexual licentiousness and indecency. In Kalyani's discussion, however, 'living together' is positively associated with individual freedom and choice while marriage is negatively understood in terms of institutional control.

At the same time, she conveys the dual sense of rebellion and disenchantment with the question of 'marriage' in the movement:

> But it also happened that people have not hesitated to divorce, many left. Have at any time said I'll divorce you ... divorced just through words [...] It's very difficult to see individual relationships and I don't want to dissect them. I just want to speak generally. You can't capture all the 'equations' of personal life.

By drawing attention to the fact that many of the alliances formed at the time did not last (our first contradictory discourse), Kalyani points to some of the limitations of a revolutionary discourse of marriage. Given the arbitrary nature of 'divorce' at the time, the freedom to experiment with interpersonal relationships did not always guarantee security, especially for peasant women. In contrast to other narratives, the concept of 'living together' is not denounced as immoral licentiousness but neither is it valorized given the heartbreak of divorce and abandonment.

In recognizing the subversive potentials of the practice of 'living together' and its concomitant failures, Kalyani's account of 'marriage' in the movement does not contradict her contemporary feminist politics or her social persona as a women's rights activist. On the contrary, by critically evaluating 'marriage' in the movement's history through a contemporary feminist lens all of Kalyani's observations complement our recognition of her as inhabiting a particular subject-position which she does throughout the interview – that of a middle-class woman, feminist-activist. Her past and present seem to be fused together to achieve subjective composure. Yet this composure rests on the weight of a studied silence that pervades her entire movement-story.

Kalyani's first relationship in the movement to which she briefly alluded in the fragment above broke. Throughout the interview, she fiercely protected her privacy with regard to it:

> [*Were you married then?*] No, in 1978. And then marriage ... there was a relationship; there was no marriage. These social customs were broken. 1975–1976 I had a relationship. I actually don't want to say anything about that, don't want to bring in my 'personal life'.

While Kalyani evaluates 'marriage' in the collective history of the movement, she chooses not to speak of her own personal history. This story of a relationship formed and broken is explicitly expelled from her life-story. Kalyani's silence is, however, a powerful one. By emphasizing (more than once) that she does not wish to disclose certain happenings, Kalyani makes this silence part of the interaction between herself and her audience, the interviewer. The silence is thus inextricably tied to the kind of social recognition that she wishes to elicit from me and is, in this way, linked to her attempts to achieve subjective composure. Self-composure, it seems, can only be achieved in the silencing or abjection of certain pasts. The abject, in Kristeva (1982), is what is expelled from the self but also constitutes the self: '[...] I abject *myself* within the same motion through which 'I' claim to establish *myself*' (1982: 3). At the same time, the constitution of the subject through abjection renders it potentially fragile and unstable insofar as what is repudiated threatens to expose the presumptions on which subjectivity is grounded, and to dissolve the subject itself.

I do not know why Kalyani chose to remain silent about her first relationship or what it is that she was unwilling to reveal. Practices of expulsion and silencing can be viewed as strategies for coping with risky, even traumatic memories that threaten self-composure (Thomson, 1994). They can also suggest a dissonance between personal and public memories. Private memories that do not 'fit' with public narratives, to recall Thomson, often merit active forgetting and repression. For Summerfield (1998, 2000), this lack of fit is particularly true for women whose stories are less likely to be returned via the cultural circuit than men's. Given that cultural discourses may not always be responsive to gendered experience, women end up 'muting' their thoughts and feelings, or altering their stories to make them conform to dominant discursive frameworks. These practices of silencing are especially pronounced when women's experiences deviate from societal norms to do with marriage, motherhood, and sexuality, for instance. Women consequently find it harder to place their memories within publicly sanctioned cultural discourses and to express themselves in familiar and shared terms. To this extent, certain past positions become unspeakable for women since they threaten social recognition and acceptability in diverging from normative constructions of femininity. The need to silence personal stories can be linked to this relation between memory and culture, that is, to the lack of fit between the public narrative repertoire and individual memories that necessitates the silencing or 'forgetting' of certain pasts.

The discordance between female subjectivity and cultural storylines of gender offers us a provisional reading of the gender politics of Kalyani's silence. Her need to silence a personal story of 'marriage' is suggestive of the limitations of available discursive repertoires (around 'revolutionary marriage') to culturally validate personal experience. Such a silence is not Kalyani's alone, and can be located in a more shared silence that structures women's stories on marriage in general.

Women's narratives often voice betrayal, disappointment and anger at the way in which everyday life remained unchanged even though they 'married' so-called political men. For most of the women I interviewed, marriage to 'political' men within the movement, as opposed to an arranged marriage, came with the promise of a radically altered organization of gender relations and the domestic sphere. Idealized revolutionary masculinity included, for them, a commitment to gender equality. Yet what these women's narratives repeatedly underscore is the disparity between expectations and reality, between ideals and their actualization. The contradiction between men's public and private lives comes out forcefully in their accounts on conjugality. While male activists were urged to renounce the role of the 'householder' for that of the 'revolutionary', women were expected to conform to middle-class norms and expectations of domesticity and womanhood.[15] 'Revolutionary marriage' within the movement was thus a site that potentially reproduced middle-class ideals of femininity, wifely submission, and sexual respectability that most women (including Kalyani) had consciously left behind. The next section demonstrates more trenchantly how beneath the apparent radicalism of the Naxalites with regard to the institution of marriage, there were manifest continuities with normative codes of sexual morality. Identification with regulatory ideals of gender, (hetero)sexuality and class was very much implicated within the radical redefinition of marriage.

While these women mark the discursive limits of 'revolutionary marriage' in the movement, their stories are eventually silenced in the public sphere, expressed in an overwhelming need to remain anonymous. Barring one, all of the women who voiced their experience of conjugal betrayal expressed a strong desire to remain anonymous like Kalyani herself. Kalyani's individual silence has to be located, I believe, in this collective silence that structures women's stories on marriage as a whole; a silence that is suggestive of the difficulties of conforming to a utopic vision of 'marriage' that nevertheless demanded adherence to hegemonic gender norms. For women like Kalyani who came to the movement for its promise of a new egalitarian society (that included gender equality), this incongruence between ideals and reality becomes all the more significant. Perhaps it is this collective betrayal and the inadequacies of our narrative registers to give it meaning that renders it unspeakable within the public domain.

Kalyani's silence is, in this way, suggestive of a more general kind of silence – the lack of the culture in which one lives to provide adequate narrative resources to meaningfully figure one's past and the anticipated future (Freeman, 2000). A

15 Ex-Naxalite and established Bengali writer, Joya Mitra (1994) writes scathingly of her first marriage (at the age of nineteen) with an older communist man. She writes of the disparities between communist ideals and everyday practices in the manner in which she was expected to fulfill her wifely duties, curb her political involvement, and support her child in the face of her husband's renunciation of household duties as relics of traditionalism. Joya's feelings of anguish that even led her to contemplate suicide resonate with the personal testimonies of some of my interviewees. These have yet to be articulated in the public domain like Joya's. The theme of betrayal runs through the representation of ro-

dominant discourse of 'revolutionary marriage' is clearly incongruent with women's own experiences of 'marriage' in the movement. While women often tell a different story of 'marriage' that voices this incongruity, they do so under the threat of discomposure, indicated, I have suggested, by their preference to remain anonymous. The marginal position that these memories continue to occupy within the cultural circuit suggests their inability to constitute an oppositional narrative that can embrace (rather than abject) the ambiguities, politics, and traumas of lived experience. (The inability to tell a story has to be located not only in the lack of narrative fit between the movement's discursive repertoire and personal memory but also in the absence of an oppositional story that can re-write the past.)

'Feminism' can be thought of as one such oppositional narrative within which women can forge new identities and find new voices (Plummer, 1995). As a women's rights activist, Kalyani is indeed invested in the narrative idioms of the contemporary woman's movement in India. Yet her silence is, I have suggested, part of her efforts to coherently constitute this very identity. Thus even the language of feminism entails, it seems, certain silences, expulsions, and 'forgettings' that mark the boundaries of what can be recollected, and what can be told. Rather than enabling hitherto silenced pasts to be spoken, renewed political subjectivity rests on the operation of abjection and repudiation.

Kalyani's active silencing of a past relationship reflects the failure of a shared discourse of revolutionary marriage to legitimately frame the dynamics of interpersonal relationships formed at the time. Above and beyond this, her silence demonstrates the fragility of all the subject positions with which we identify, even those that have only been recently made available by feminism. Subject positions always entail the loss of alternative identifications that nevertheless 'continue to contain these within, to refer to them, and be disrupted by their liminal presence' (Redman, 1999: 75). This perpetual threat of discomposure raises the political question of the cost of articulating a coherent identity through abjection, and of who bears this cost, this burden of silence. In the narrative that follows, similar strategies of repudiation ground narrative coherence and identity.

mance and interpersonal relationships in filmic narratives of Naxalbari as well. In Buddhadeb Dasgupta's *Dooratwa* (The Distance, 1978), Anjali marries Mandar, a former revolutionary, to provide an identity for the illegitimate child she is carrying, and also because she believes he is 'different'. Unable to accept her premarital affair, Mandar, however, rejects her. A recent film by Satarupa Sanyal, *Anu* (1998) tells the story of a woman (Anu) who awaits the release of her Naxalite lover from prison. On his return, the 'husband' rejects her on learning of her rape by political goons, signaling, once again, the divergence between the ideals of gender equality, and entrenched middle-class patriarchal values.

respectable 'revolutionary' sexuality

Saumen was a middle-ranking activist of the Naxalbari movement whose wife was a sympathizer. His narrative is one of the only thoroughgoing male discussions of 'marriage', and more significantly, the only one that fully develops a problematic alluded to by Kalyani, which is the dovetailing of 'marriage' with sexual violence.

In common with other narratives (and in the popular memory of the movement), Saumen's narrative represents the Naxalite subject as a specifically male one, and infuses it with certain qualities. Women are represented as sympathizers who took several 'risks' for their revolutionary partners. Both Saumen and his wife, Latika, suggest that women saw young male Naxalites as their 'spokesmen' who could voice their shared sense of oppression in the face of their own collective inability to do so. In the Naxalite, the archetypical woman-in-love sees the possibility of rupture, change and renewal that she cannot herself affect. In this imagining of an idealized revolutionary masculinity lay a particular vision of dependent femininity, not very different from the way in which women have been represented in elite nationalist politics as the husband's helpmate. This construction of revolutionary masculinity and, correspondingly, of a dependent femininity echoes literary representations of the time. For instance, in *Antarghat*, the male Naxalite tells the female protagonist, 'Do you want to die with me, Bibi?' She replies: 'If that is my destiny, then yes, I do.' Similarly, in *Manabputri*, the revolutionary tells his newly wed wife that even Lord Ram could not leave his wife behind and go into exile. Thus she too will accompany him into the villages for political work.

In these narratives romance is represented overwhelmingly in gendered terms, with women positioned as the receivers and men as the initiators of romance. At the same time, these romantic texts are framed within an overriding narrative of rupture – that of revolution and change. To this extent, they seem to erase power differentials to emphasize union and mutuality in aid of revolution. Thus both the man and the woman dream the same dream of a better world, and Bibi (in *Antarghat*) is willing to die *with* Antu for the sake of 'revolution'. Similarly, in much of the poetry of the movement (like the poem we began with; see also Jalark, 1998) where *bhalobasha* is a dominant theme, there is an evocation of 'we', suggestive of the unity and equality of the male-female partnership forged through a common revolutionary goal. Framed within this larger discourse of revolution, romantic love does not have one particular fixed meaning. Thus, although femininity and masculinity come to acquire fixed essences in Saumen's narrative, there are moments in his discussion that open up the possibility of reading romance *in revolution* differently. Thus he says (on behalf of the woman-in-love): 'I am not looking for a husband; I am looking for a comrade', to emphasize a reversal of the gender hierarchy that structures traditional conjugal relations. This is, however, a lost opportunity. The overriding ideology of Saumen's narrative is that of dependent femininity and heroic masculinity that, in fact, brings into question the entire rhetoric of 'comrade'.

Latika and Saumen did have a truly 'revolutionary' marriage. It was neither a religious nor a registered wedding; they come from different familial and caste backgrounds (he is a Kayastha caste while she is Brahmin),[16] and she is also at least ten years older than him. Saumen calls their wedding 'absurd' since neither his in-laws nor Latika herself knew his real name at the time of marriage; they only knew him by his pet name. It is important to underscore at this point that

16 Kayastha is a dominant caste group that has a significant degree of economic and social

this story of his marriage to Latika was one that I heard more than once, and is one from which Saumen derives (I believe) a certain psychic comfort given his traumatic life in and after the movement. In many ways, this story is crucial to his achievement of narrative composure. In the extract that follows, we see how Saumen negotiates recognition and composure with contradictory constructions of revolutionary marriage as sexual licentiousness (our second contestatory discourse). Such a discourse on sexual liberalism renders his subject-position less stable and threatens to produce discomposure. It is also in this discussion that an overwhelming concern with sexual respectability shapes the revolutionary discourse on marriage.

capital but perhaps not as much cultural capital as the Brahmins. They comprise of a large section of the Bengali middle-class, and are usually educated, affluent, and in white-collar employment. The marriage of an upper-caste (Brahmin) girl with a (Kayastha) boy is, however, not normative, and defies the codes of hypergamy.

Saumen's discussion took a radical turn when I asked him if he thought that the movement was progressive or liberal in terms of relationships. His response makes clear his understanding of my question as one of sexual licentiousness in the Party.

> In terms of relationships, if there is any question of 'sexual freedom', they are certain 'foreign elements' or you can say 'deviations'. That's how we saw it. This was not the usual matter. Some people brought it in. What I am saying as 'liberal' is something else. A boy likes a girl; the 'variables' that usually enter into this liking were absent here. But now what is being said is being forced on the Naxalite movement, that 'live together', 'pre-marital' this and that. But even in our time, these were there but were thought of as outside elements, as 'deviations', 'liberty'.

Latika interrupts to say that these were not a fallout of the movement and Saumen continues:

> Absolutely. These have been made up. A lot of people have done this. You won't even see this in the literature, what today, what in the name of 'feminism' or something else, what, you know, various types of 'free mixing' – this didn't exist in our time. But obviously some girls and boys got the opportunity to 'freely live' and work in one area, but deviations happen; there's no point theorizing them. They are just deviations. For instance, at Bankura we saw an 'apparent' closeness between a girl and a boy. I mean a real closeness that was *acceptable* to us. So we all sat and as a gang told them 'get married'. It was just the opposite actually. *We never let anyone take liberties*. It was just the opposite. We didn't like it at all. [...] We would always think that we shouldn't get a bad name. [...] (Emphasis added).

Much of the above extract works in order to convince me, the audience, of a particular definition of the word 'liberal'. Saumen sharply rejects 'liberal' as implying 'sexual freedom' or 'living together' from his own usage of the word to describe the privileging of love and companionate marriage in the movement. While Kalyani differentiates between 'marriage' and 'living together' to assert that individuals were *not* married in the Naxalite movement, Saumen vociferously argues 'just the opposite'. The concept of 'living together' insofar as it is socially unacceptable to the *bhadralok* class is rejected as a 'deviation' from the norm 'marriage'. 'Liberal' for him means a rupture in the traditional discourse of arranged marriage where love is subordinated to other 'variables' such as economic status, social class and caste. Yet 'liberal', he repeatedly reminds us, did not translate into an acceptance of sexual licentiousness in the movement. He condemns any form of sexual liberalism in the strongest of moral terms, betraying his rootedness in a *bhadralok* discourse of respectability, of which sexuality is a crucial signifier. Such a discourse hinges on the sexual modesty of women, the sacrosanct markers of respectability for the Bengali middle-class. Interpersonal relationships in the movement had to fit such an 'acceptable' standard of sexual morality, and were consequently subject to moral censure and public policing. In paradoxical ways, Saumen's narratives suggest that while everything changed in terms of conjugal relations in the movement, everything actually remained the same. While 'marriage' in the Naxalbari movement might have upset established hierarchies of caste, class and age, it seems to have been reticent on issues of gender and sexuality.

Saumen's discussion also evokes a more shared, collective discourse on sexuality that can be recognized as the Party's code of sexual conduct. He says: 'And generally, as a rule, if a Party boy was close to a girl then it was taken for granted that that was his wife'. Within such a discourse on sexuality relationships could not but end in marriage, and the question of individual consent often becomes a tricky one, especially for women. Narratives on domestic abuse speak of some of the implications of the Party's regulatory stand on sexuality. The idioms of sexual licentiousness within the movement were also gendered insofar as it was women who were, as the bearers of public morality, the objects of policing and slander. Once again, this made 'marriage' mandatory for women.

Although much of Saumen's narrative attempts to deny any kind of promiscuity in the movement, it is less than consistent. The threat of discomposure occurs precisely when Saumen is forced to accept the existence of licentious sexual behaviour that points, moreover, to the discursive limitations of 'revolutionary marriage'. These sexual possibilities are immediately disavowed through repudiation and expulsion, even as they are identified. Thus, even when Saumen does admit the prevalence of 'sexual freedom', he identifies it as a 'foreign element', a 'deviation', something that was not the norm and certainly not tolerated. Here Saumen's language works in particular ways to render sexually unacceptable behaviour as something that was not intrinsic to the movement; it was not a fallout of the movement, as he and his wife repeatedly told me. Some of his language is abstract and allusive as the language of sexuality often is in the narratives of middle-class Indians (see Puri, 1999). The possibility of a precarious sexual situation is only hinted at by saying 'pre-marital this and that', 'loose behaviour' or 'what in the name of feminism'. What is more consistent in Saumen's speech is his use of the English language to describe a pattern of behaviour that is unacceptable to him. Repeatedly, he uses phrases/words like 'free mixing', 'live together', 'girlfriend', 'pre-marital', 'liberties' and so on, to define what 'didn't exist' at the time, or what 'we didn't like at all'. His usage of English (rather than Bengali) to describe such forms of socially unacceptable behaviour is indicative of the particular position that he takes up in this discourse of sexuality. The use of English creates the possibility of rejecting certain patterns of social behaviour as absolutely alien, other and external to a specifically Bengali sensibility. Such forms of behaviour are banished outside the subject's own moral universe and attributed to an external agent. His use of English suggests that such an external agent could be 'westernization', a common culprit of the changing codes of sexual conduct in India (see Basu, 2001; Kapur, 2005). Saumen, in fact, pointedly evokes 'feminism' to explain the existence of 'various types of free-mixing' in contemporary times. It would be fair to assume that feminism, for Saumen, is closely related to the project of 'westernization' and 'modernization' in general, and not perceived as something rooted in an Indian cultural context. My point here is that it is through the use of English that Saumen disavows certain sexual possibilities, renders them external to and outside the domains of community. In this way, Saumen composes a certain sense of what it means to be Bengali through his use of English.

As with Kalyani, potential discomposure in Saumen is negotiated through acts of repudiation, disavowal and exteriorization. The possibility of non-normative sexual practices can be accepted only by discursively rendering them *outside* the 'movement community'. By exteriorizing such acts, narrative practices preserve a certain version of the community and the self. Here the conception of community that most affords Saumen with psychic comfort is that of a morally and socially acceptable (and respectable) entity. The possibility of sexual licentiousness in the movement does more than simply hurt his moral sensibilities. It disrupts the social recognition that his own *individual* story of a revolutionary marriage affords him by questioning its legitimacy. At the same time, it crucially threatens a *collective* identity that is formed in the (male) image of the self-righteous revolutionary martyr, a key icon of Naxalbari. Revolutionary masculinity can only preserve itself through the abjection of those discourses that threaten its coherency such as those of alternative masculinities, sexual moralities, and elsewhere, of male sexual power.[17]

As a final point, I turn to Saumen's discussion of the sexual exploitation of peasant women by middle-class 'revolutionaries'. *Bhadralok* comrades, he says, often 'married' peasant women as testimony to their complete integration and 'de-classed' nature:

> The boys who came from Presidency College (I won't give you any names but what I'm telling you is 'authentic') – many good, well-known Presidency College boys with 'good results', who went to Gopiballavpur, their 'method of integration' was to stay at peasant homes as man and wife. And under the influence of some famous Naxalite leaders, young unmarried girls got 'pregnant' there, peasant

17 The past that is repudiated is not only one of sexual licentiousness. The imagining of revolution and its heroes is founded upon a collective denial of male sexual violence. This is evident in the collective reticence to discuss the problematic of sexual violence in the movement, especially the violence faced by women comrades *within* the revolutionary domain. The memory of sexual violence is, as a whole, one that is repeatedly repudiated for the sake of preserving the cultural identity of the Naxalite built around the image of the revolutionary martyr.

> girls. I can bring you a witness who has 'helped' one such girl. They would go and as man and wife, would be with this girl and that girl [...] then at another place, they would also be husband-wife. [...] This incident, this 'reckless' behaviour that I can do anything and that it is 'justified' in the name of revolution [...]

In the case of peasant women, a blurring of the boundaries between 'marriage' and sexual abuse occurs within a masculine discourse of becoming 'de-classed'. Subaltern female sexuality functions, in this instance, as a metonym for a truly de-classed revolutionary masculinity. Saumen goes further to suggest that the radicalization of marriage practices that took place under the rubric of a revolutionary politics provided a legitimate framework for sexual violence. Acts of male sexual violence remained couched within the fantasy of 'revolution' that promised a class-based societal transformation. The vulnerability of subaltern women to male sexual power remained embedded, routinized, and normalized within this vision of a transformative class politics.

In implicating the 'brilliant students' of Presidency College, the premiere college of Kolkata, and not *all* male Naxalites for acts of sexual violence, Saumen's narrative is able to preserve a coherent self-image (self-righteous, respectable, revolutionary) in the face of a threatening counter-memory that could produce discomposure. His condemnation of elite Presidency College students preserves true revolutionary subjectivity in the image of the lower middle-class, lower-caste (male) political activist. Throughout his movement-story, Saumen ascribes authentic revolutionary subjectivity to this community of male activists (to which he belongs) rather than to the elite intelligentsia who have been the chief beneficiaries of public recognition. No doubt, this (re)imagining of the revolutionary subject serves to stabilize his own place in history, and to compose self-identity in its image.

conclusion

In the narratives visited here the achievement of 'composure' is that of coming to grips with contradictory and competing constructions of the past. Marriage-stories are also, we have seen, a part of the narrative construction of identity; individuals invest in particular subject-positions that offer them a degree of psychic comfort. Such 'acceptable' self-images include that of Kalyani's as a middle-class feminist-activist, and that of Saumen's as a middle-class Bengali man with strong political (and moral) commitments. They also include discursive images of revolutionary masculinity and dependent femininity or even sexually respectable femininity. Individual imaginings of masculinity/femininity find resonance in more popular forms of memory, underscoring the mutual dependence between stories of the self and the public narrative repertoire of a culture (Redman, 1999). Kalyani and Saumen draw on the narrative resources of a lived Bengali middle-class culture to offer a particular view on romance and sexuality that echo the values of the wider group to which they belong. In doing so, they are able to elicit social recognition in ways that aid the achievement of composure.

Yet I have found that 'composure' is founded on an 'abject zone' that threatens to undo the coherent stories that we tell about ourselves, and the forms of identity in which we take shelter. Kalyani's silence is one such moment in her life-story that consolidates self-identity through the repudiation of certain pasts. In Saumen, the idea of a coherent unified subject, rooted in a discourse of revolutionary manhood, is threatened when confronted with a contradictory discourse of 'liberal' sexuality. In both Kalyani and Saumen, the achievement of composure is fragile insofar as what is repudiated and disavowed gets (re)inscribed on the surface of identity – '[it] lies there, quite close, but it cannot be assimilated.' (Kristeva, 1982: 1). Kalyani's projection of a 'core' authentic self *requires* her silence, just as Saumen's appeal to revolutionary masculinity incorporates the repudiation of illicit sexuality.

It is particularly revealing that 'marriage' constitutes such an abject zone, for women *and* men in the history of Naxalbari. Through the narratives visited here we have seen how, as in the case of other revolutionary movements, a utopian vision of revolutionary marriage conformed to the received conventionality with regard to gender and sexuality. It demanded adherence to hegemonic gender norms and *bhadralok* sexual moralities that replicated, moreover, the sexual taboos that existed outside the revolutionary domain within it. In the final instance, it is this disparity between the ideal and the actual that makes 'marriage' such a contestatory site in the cultural memory of the movement, and demands, in turn, the abjection of certain risky memories. Both Saumen and Kalyani's narratives on 'marriage' are composed through a repudiation of those pasts that acquire a liminal status within shared (and hegemonic) norms of marriage and sexual morality, both within the revolutionary domain and outside of it. Their narratives are exemplary for what they suggest about the difficulties of conforming to a radical social practice that nevertheless mirrored the reality it attempted to disrupt and radicalize.

acknowledgements

I thank the *Feminist Review Trust* for its contribution towards the funding of this project, and the anonymous referees at *Feminist Review* for their fruitful engagement with this article. My thanks also to Deborah Steinberg, Parita Mukta, Gauri Raje and Rafael Winkler for their comments on various versions of this piece.

author biography

Srila Roy is a doctoral candidate at the Department of Sociology, University of Warwick where she teaches on gender/sociology. Her Ph.D. research engages with issues of gender, violence and memory in relation to the 1970s extreme Left, Naxalbari movement of West Bengal.

references

Banerjee, S. (1984) 'India's simmering revolution: the Naxalite uprising', London: Zed.

Banerjee, S. (1987) editor, *Thema Book of Naxalite Poetry*, Calcutta: Thema.

Basu, B. (2002) *The Enemy Within*, Translated by Jayanti Datta, Delhi: Orient Longman.

Basu, S. (2001) 'The blunt cutting-edge: the construction of sexuality in the Bengali 'feminist' magazine *Sananda' Feminist Media Studies*, Vol. 1, No. 2: 179–196.

Byrne, B. (2003) 'Reciting the self: narrative representations of the self in qualitative interviews' *Feminist Theory*, Vol. 4, No. 1: 29–49.

Dasdhikari, S. (1998) editors, *Shotorer Shahid Lekhok Shilpi*, Kolkata: Adhuna Jalarka Parishad.

Dasgupta, R. (2003) *Marxism and the Middle Class Intelligentsia: Culture and Politics in Bengal 1920s–1950s*, Unpublished D.Phil Thesis, Oxford University.

Dawson, G. (1994) *Soldier Heroes: British Adventure, Empire and the Imagining Of Masculinity*, London: Routledge.

Devi, M. (1997) *Mother of 1084* Translated by Samik Bandyopadhyay, Calcutta: Seagull Books.

Engels, F. (1884) *The Origin of the Family, Private Property and the State*, New York: International Publishers, 1972 edition.

Evans, H. (1992) 'Monogamy and female sexuality in the People's Republic of China' in Rai, S., Pilkington, H. and Phizacklea, A., editors, *Women in the face of change: the Soviet Union, Eastern Europe and China*, London: Routledge.

Freeman, M. (2000) 'When the story's over: narrative foreclosure and the possibility of self-renewal' in Andrews, M., Day Sclater, S., Squire, C. and Treachers, A. (2000) editors, *Lines of Narrative: Psychosocial Perspectives*, London: Routledge.

Gluck, S.B. and Patai, D. (1991) editors, *Women's Words: The Feminist Practice of Oral History*, New York, London: Routledge.

Hollway, W. and Jefferson, T. (2000) *Doing Qualitative Research Differently: Free Association, Narrative and the Interview Method*, London: Sage.

Johnson, Richard (1982) 'Popular memory: theory, politics, method' in Johnson, R., Sutton, D. and McLennan, G., editors, *Making Histories: Studies in History-writing and Politics*, London: Hutchinson in association with Centre for Contemporary Cultural Studies, University of Birmingham.

Kannabiran, V. and Lalitha, K. (1989) 'That magic time: women in the Telengana people's struggle' in Sangari, K. and Vaid, S., editors, *Recasting Women: Essays in Colonial History*, Delhi: Kali.

Kapur, R. (2005) *Erotic Justice: Law and the New Politics Of Postcolonialism*, London; Portland, OR: Glass House Press.

Kristeva, J. (1982) *Powers of Horror: An Essay on Abjection*, Translated by Roudiez, L.S., New York: Columbia University Press.

Lahiri, A. (2001) *Postwar Revolt of the Rural Poor in Bengal*, Calcutta: Seagull Books.

Lawler, S. (1999) 'Getting out and getting away: women's narratives of class mobility' *Feminist Review* No. 63 Autumn: 3–24.

Lawler, S. (2002) 'Narrative in social research' in May, T., editor, *Qualitative Research in Action*, London: Sage.

Mitra, J. (1994) 'Cholte cholte' in Mandal, P. and Mitra, J., editors, Shei Doshok, Kolkata: Papyrus.

Mitra, S. (1993) *Manabputri*, Kolkata: Dey's Publishing.

Passerini, L., Selma, L. and Paul, T. (1996) editors, *Gender and Memory*, Oxford: Oxford University Press.

Plummer, K. (1995) *Telling Sexual Stories: Power, Change and Social Worlds*, London: Routledge.

Poldervaart, S. (2000) *The Recurring Movements of 'Free Love'* Written for the workshop 'Free Love and the Labour Movement', International Institute of Social History, 6 October 2000. Online. http://www.neha.nl/~womhist/polder.pdf.

International Institute of Social History Amsterdam http://www.neha.nl/~womhist/polder.pdf, accessed 6 October 2000.

Puri, J. (1999) *Woman, Body, Desire in Post-Colonial India: Narratives of Gender and Sexuality*, London: Routledge.

Ray, R. (1988) *The Naxalites and Their Ideology*, Delhi, Oxford: Oxford University Press.

Ray, R. (1999) *Fields of Protest: Women's Movements in India*, New Delhi: Kali for Women.

Ray, R. (2000) 'Masculinity, femininity, and servitude: domestic workers in Calcutta in the late twentieth century' *Feminist Studies*, Vol. 26, No. 3: 691–718.

Redman, P. (1999) *Boys in love: narrative, identity and the production of heterosexual masculinities* unpublished PhD Thesis, University of Birmingham, Birmingham.

Ricoeur, P. (1991) 'Narrative identity' in Wood, D., editor, *On Paul Ricoeur: Narrative and Interpretation*, London: Routledge.

Sen, K. (2001a) 'Paschim banglar naxal andolan-e meyera' in Chattopadhyay, M., editor, *Esho Mukto Koro: Nareer Adhikar o Adhikar Andolan Bishayak Prabandha Sankalan*, Kolkata: People's Book Society.

Sen, M. (2001b) *In Search of Freedom: An Unfinished Journey* Translated by Stree, Calcutta: Street.

Somers, M. (1994) 'The narrative constitution of identity: a relational and network approach' *Theory and Society*, Vol. 23, No. 5: 605–649.

Summerfield, P. (1998) *Reconstructing Women's Wartime Lives: Discourse and Subjectivity in Oral Histories of the Second World War*, Manchester: Manchester University Press.

Summerfield, P. (2000) 'Dis/composing the subject: intersubjectivities in oral history' in Cosslett, T., Lury, C. and Summerfield, P., editors, *Feminism and Autobiography: Texts, Theories, Methods*, London: Routledge.

Taylor, B. (1983) *Eve and the New Jerusalem: Socialism and Feminism in the Nineteenth Century*, London: Virago.

Thomson, A. (1994) *Anzac Memories: Living With The Legend*, Melbourne, Oxford: Oxford University Press.

Tyler, M. (1977) *My Years in an Indian Prison*, London: Gollancz.

Vindhya, U. (1990) 'The Srikakulam movement' in Sen, I., editor, *A Space Within the Struggle*, New Delhi: Kali.

Vindhya, U. (2000) 'Comrades-in-arms: sexuality and identity in the contemporary revolutionary movement in Andhra Pradesh and the legacy of Chalam' in John, M.E. and Nair, J., editors, *A Question of Silence? The Sexual Economies of Modern India*, London: Zed.

Filmography:

Anu, Dir. Satarupa Sanyal, 1998.

Dooratwa, Dir. Buddhadeb Dasgupta, 1978.

doi:10.1057/palgrave.fr.9400283

83 dialogue
making political hay of sex and slavery: Kansas conservatism, feminism and the global regulation of sexual moralities

Almas Sayeed

When I returned to my home state of Kansas in the fall of 2004, there was palpable excitement among the state's small liberal community about the possibility of unseating the incumbent president, George W. Bush, in the then upcoming national election. Although few in number, the state's liberal activists had frequently found themselves on the front lines – and even occasionally on the winning side – of recent skirmishes in the 'culture wars', the battle over the regulation of sexuality, including abortion and gay marriage initiatives. For liberals in the US and, in particular, for feminists, the 'culture wars' have come to represent a high stakes battle over the appropriation of the language of 'moral values', a struggle that bears material consequences in the form of regulatory policies and laws, the brunt of which fall on women's bodies. Liberal activists in Kansas, by virtue of their minority status, clearly understood the power accorded to those who set the terms of debate. Now they hoped their hard work would pay off with a political and ideological shift in the White House, if not in Kansas.

For me it was difficult to muster much excitement given the circumstances. Modern Kansas is a sparsely populated, largely white, rural state in the centre of the US that has become a point of intersection for some of the most zealous and religiously conservative elements that exist in the country. In particular, Kansas has become known for a religious zealotry that identifies the regulation of sexuality as its number one enforcement mechanism. Operation Save America, a radical anti-abortion group linked to violence perpetrated against reproductive health clinics, staged its greatest public offensive against abortion in 1991, rallying thirty thousand testifying anti-abortionists to the city of Wichita, Kansas (where I grew up), blocking

(119–131) © 2006 Feminist Review. 0141-7789/06 $30 www.feminist-review.com

patients' access to the city's only clinic and effectively disrupting its operations. The state legislature has shadowed this fervent mobilization, implementing creative and insidious strategies to block poor women and adolescents' access to reproductive health care. Kansas is also home to the Westboro Baptist Church of Topeka headed by Fred Phelps, the infamous preacher who travels the nation with his followers to protest events ranging from 'gay pride' marches to funerals of gay activists to theatre performances, proclaiming the news that 'God Hates Fags' and 'AIDS kills fags dead'.[1] Here, ardent Catholic leaders declared the socially conservative Pope John Paul II to be 'too socially permissive', and have created an alternative Catholic Church where they appointed a more 'true-to-the-book' pope (Palmer, 1982; Farney, 1995; Frank, 2004). In Kansas, disgruntled white men, fuelled by their hatred of the government and manipulating people's fear of Islamic terrorism after the first bombing of the World Trade Center, planned their attack on a Oklahoma City government building that took 168 lives. And perhaps most famously, Kansas earned international mockery as the place where Christian conservatives on the State Board of Education removed evolution from the required science curriculum.[2] Kansas is a state where regulations of sexuality and political processes have converged to stifle opposition and debate. It is not a state in which spirited campaigning tips elections to the left: Kansas has supported the Republican ticket in every presidential election since 1964, often by more than 70% of the popular vote.

1 For more on the Phelps' Ministry see www.godhatesfags.com.

2 Thomas Frank (2004) outlines the way in which these conservative elements have converged in Kansas, emphasizing the state's class politics as explanation for the state's ideological shift to the right. In this paper, I explore the gendered and racialized components of this shift.

Yet, with the advent of the Bush administration Kansas conservatism has gone from being the butt of jokes on late night talk shows to reflecting the national political sentiment, from the fringes to the mainstream. Because of its geographic centrality, Kansas has become a nexus for those in the US seeking to strategically influence the regulation of sexuality across the country and around the globe. The backlash talk of 'moral depravity' that we, as Kansans, were accustomed to hearing from conservative leaders was now echoed by the White House. For liberal activists working to oppose the conservative movement in Kansas, as in other states, hope for an end to the Bush era was short-lived and the administration's re-election initially crippling. Bush's re-election seemed to be a stamp of approval for the administration's manipulative use of fear to justify military expansion, its backing of religious conservatism, anti-abortion initiatives and its promise to pass a Federal Marriage Amendment designed to prohibit the legalization of gay marriage. The conservative movement's current traction with the White House has propelled it to take on new issues, flying under the twin banners of 'moral repugnance' and 'freedom fighting abolitionism' in pursuit of moral hegemony. The success of this movement can, in part, be attributed to its strategic and slippery appropriation of the language of 'moral values'. Feminist analysis of these rhetorical devices is integral to formulating a response.

In this piece, I explore the connections between Kansas and national conservatives, emphasizing the importance of place and location in mapping connections between political processes and structures of power. I examine the way conservatives have generated 'moral repugnance' towards three issues crucial to the regulation of sexual morality: abortion, gay marriage and international sex trafficking. Although these issues are not the only foci of the conservative movement – for example the occupation of Iraq is of vital interest here and worthy of further exploration – these issues offer interesting, even overlapping areas of intersection for feminists. Finally, I consider what this backlash, as waged by conservatives via the 'culture wars', means for US feminist political engagement and for building a transnational feminist coalition.

bleeding Kansas: moral repugnance, abolitionists and the politics of abortion

As one of the strongholds of the conservative movement, the anti-abortion movement in Kansas has implemented some of the nation's steepest obstacles to women's abortion access, particularly with respect to poor women and adolescents. With control of the White House, conservatives have had success generating 'moral disgust' towards abortion by strategically positioning a woman's 'right' to choose as analogous to a plantation owner's 'right' to possess human slaves. Positioning the unborn foetus as comparable to the captured black slave has been a successful discursive move for the anti-abortion movement in Kansas, both politically and symbolically. This metaphor has particular resonance in the state as Kansas was the pre-Civil War border between Free and Slave states and hosted some of the war's bloodiest battles, earning it the name 'Bleeding Kansas'. To this end, ardent conservatives of the Kansas variety were the first to compare Roe v. Wade, the 1973 Supreme Court case that legalized abortion, to the Supreme Court's Dred Scott decision in 1857, which struck down state laws banning slavery within their boundaries (Frank, 2004: 183–188). Although the anti-abortionists' comparison of their movement to the work of abolitionists is emphasized in Kansas, it is not unique. Current leaders of the national movement often repeat the analogy and have emphasized that, like slavery, every means should be used to overturn abortion. So while many people were confused when Bush referenced the Dred Scott decision in the 2004 debates, conservatives knew he was speaking directly to them. As The Nation's Katha Pollit notes, 'For Bush, Roe=Dred. Spread the word' (Pollit, 2004).[3]

3 An important point of note here is that although the Supreme Court's *Dred Scott* case is often used to illustrate the work of 'activist judges,' – those who bend laws

At a strategic level, the state's conservative leaders have played a vital role in keeping the issue of abortion at the forefront of national public discussion. Although former US Attorney General John Ashcroft made international headlines in 2004 by dispatching federal marshals with subpoenas to clinics around the nation in order to confiscate women's medical records (Lerner, 2004; Associated

Press, 2004), Kansas' current Attorney General Phil Kline had been quietly trying to access the files of women who had sought procedures considered illegal in Kansas since 2003 (Sylvester, 2005). In 2005, Kline placed Kansas in the spotlight of the national abortion debate through his campaign to 'crack down' on child sexual abusers and remove their privilege of 'secrecy' by subpoenaing clinics for the medical records of adolescents who had undergone abortions (Jones, 2005). In a legal opinion, Kline argued that mandatory reporting laws require medical professionals to report all physical and sexual abuse of a child to government authorities. A minor's pregnancy, Kline stated, would be clear evidence of such abuse and the state should step in if a doctor performs an abortion on an adolescent girl without informing authorities about the procedure.

to match their personal political will – in fact, the decision strictly adhered to accepted definitions of 'property,' 'slave' and favoured states' rights. This was very much in line with the political sentiment of the time and is not, in this regard, a particularly good comparison to *Roe v. Wade*.

While conservatives' campaign to obtain patients' medical records will likely prove unsuccessful in the courts, Kline's major victory lies in the publicity surrounding the language of his campaign. Adopting the rhetoric of 'late-term' and 'partial-birth' abortion – terms that suggest the foetus has attained viability – and claiming to crack down on 'child abusers', Kline successfully generated both broad moral disapproval towards abortion and developed effective strategies to influence the state legislature to restrict women's access to reproductive health care. In this vein, the conservative movement's language of righteousness and moral urgency is used as ably to support violence in one form as to condemn it in another. The website of Operation Save America declares, 'Like slavery before it, abortion is pre-eminently a Gospel issue requiring any means to stop it'.[4]

4 See www.operationsaveamerica.org.

Because issues of privacy and doctor–patient privilege have historically framed discussions around abortion, the US pro-choice movement has struggled to maintain the focus on women's autonomy (Fried, 2002). The resurgence of religious conservatism in places like Kansas, however, has prompted a discursive shift in the pro-choice movement. The Democratic Party, historically a staunch ally of the pro-choice movement, has made a strategic retreat and is now beginning to embrace the discourse of 'moral repugnance'. In an effort to realign its centre to reflect the new Middle America, Democrats proclaim that abortion procedures should be rare and that the terms 'choice' and 'abortion' should be ideally struck from the Democrats' lexicon.[5] Although implying that women's access to health care services and contraception must be improved, the new party dogma rarely makes this explicit. Instead, many Democrats now highlight the importance of reducing the number of abortions.

5 Comments from Howard Dean, chair of the Democratic Party, on the political talk show, 'Meet the Press with Tim Russert,' NBC, News, May 22nd 2005 reprinted via transcript, available online at www.msnbc.masn.com/id/17924139/.

The anti-choice movement's success in capturing the discourse of 'morality' and its leaders' self-declared position as 'abolitionists' has quietly made abortion an issue of strategic negotiation for the pro-choice movement, often sacrificing poor women's and adolescents' reproductive rights in order to keep Roe v. Wade from being overturned. In 2004, Kansas passed a law that effectively restricts women dependent on state medical assistance from seeking an abortion by passing legislation that prohibits the state from subsidizing the procedure except in cases where the life of the mother is at high risk or the pregnancy is the result of rape or incest.[6]

6 Kansas Medical Assistance Program Professional Services Provider Manual, (January 2004), at http://www.kmap-state-ks.us/public/providermanuals.asp.

Similarly, in 2005 the Kansas legislature passed the 'Child Rape Protect Act' (Supplement to Preliminary Summary of Legislation, 2005), which requires abortion providers to preserve foetal tissue from procedures performed on girls under the age of 14. The tissue is then labelled with the patient's name and address, as well as the names and addresses of the girl's guardians, and sent to the Kansas Bureau of Investigation for DNA testing followed by a criminal investigation of child sexual abuse, regardless of the nature of the pregnancy. A consensual relationship and a child rape case are treated the same way. Because there are currently no banks of male genetic material in the country, the legislation places the onus to name the father on the adolescent girl, undermining her privacy and, in some cases, putting her in further jeopardy. Faced with such initiatives, the pro-choice movement has picked its battles carefully, choosing to limit its mobilization in battles it believes cannot be won.

The effects of the anti-abortion movement are felt well outside of the US's conservative centre. In early 2005, Bush, galvanized by members of his conservative base like Kansas Senator Sam Brownback, reinstated the 'Mexico City Policy'. This regulatory policy, also known as the 'global gag rule', explicitly bars US foreign aid funding for international development and non-governmental organizations that either give counselling or referrals for abortions or lobby against their governments' restrictive abortion laws. Contrary to published USAID reports, the Center for Reproductive Rights reports that the gag rule has had a detrimental impact on women's global reproductive health access. Initially reinstated during Bush's first administration in 2000, by 2002 the gag rule resulted in an end to all shipments of USAID-donated contraceptives to 16 countries and ended aid to leading family planning agencies in another 13 countries.[7] In addition, of the 56 countries receiving USAID assistance, 35 technically permit abortion without restriction or under particular circumstances. The gag rule forbids local NGOs in these countries who depend on USAID assistance from providing abortion related services, even if they use non-USAID funds to provide counselling or safe and legal abortions. In support of the policy, Brownback stated, 'This is taxpayers' dollars used to support organizations supporting abortion overseas. Clearly 70 plus percent of the American public would be against that' (Allen, 2005). The effects of the administration's anti-choice policies continue to chip away at women's access to reproductive healthcare both at the state level in the US and abroad.

7 'Myths and Realities: Debunking USAID's Analysis of the Global Gag Rule' Published by the *Center for Reproductive Rights* (2005), available online at http://www.crlp.org/hill_int_ggr_myths.html. Also see 'Breaking the Silence: The Global Gag Rule's Impact on Unsafe Abortion,' a report published by the Center for Reproductive Rights (2005).

queers, bigots and family values...oh my: gay marriage bans and electoral politics

The momentum of the conservative machine on the issue of abortion pales in comparison to the fervour generated over homosexuality and gay marriage. The 2004 national election confirmed conservatives' commitment to policing the sexual habits of consenting adults as gay marriage became a central focus of the nation's 'culture wars'. Campaigning as a champion of 'traditional values', Bush gained a broad base of support by once again appropriating the language of 'moral righteousness'. Not only did the Bush campaign, and the conservative movement supporting him, commandeer the language of morality in fighting for 'tradition', 'family', and 'the sanctity of marriage', but it also claimed these 'values' and 'way of life' were 'under attack' by supporters of gay rights. The language of fear and call to battle created a sense of urgency on the issue despite the fact that gay marriage is illegal throughout much of the country. The urgency was not even drawn from the possibility of legalization through a win for the opposing party; Democratic presidential candidate, John Kerry, while articulating support for civil unions, also opposed legalizing gay marriage. Although Kerry's position demonstrated enough tolerance of homosexuality to add further momentum to the conservative movement, it isolated lesbian and gay voters. Given these facts, one might ask what happened to propel the issue of gay marriage to the forefront of an already deeply polarized presidential campaign.

Part of the answer can be found in what political commentators have written about already: when Kerry's home state of Massachusetts made a move to legalize gay marriage in the run up to the 2004 election, conservatives became alarmed. If the Massachusetts law went into effect, many conservatives argued that the 'Full Faith and Credit Clause' of the Constitution, which creates uniformity on certain issues such as marriage, would have required all states to recognize those same sex-couples married in Massachusetts. In support of a Federal Marriage Amendment, the Bush campaign argued that due to the debate in Massachusetts conservatives had no choice but to push for a constitutional amendment defining 'marriage' as between one man and one woman, leaving the issue of civil unions up to the states. The possibility of this showdown prompted many to see the 2004 election as a 'national referendum on gay marriage' – one that Kerry would assuredly lose (Kurtz, 2004: 3). Throughout the debate, US feminists, both spokeswomen and organizations, remained largely silent.

The appropriation of the discourse of 'tradition' and the 'sanctity of marriage' worked with resounding success in Kansas, a place politicians tout as the barometer of conservative national values. In Kansas' 3rd Congressional District, (where I happen to live), the Republican candidate for the US House of Representatives, law professor, Kris Kobach, was positioned in the polls to take the traditionally Democratic district by storm. Running largely on an anti-gay marriage, anti-immigration and anti-taxes campaign, Kobach rallied conservatives around the claim that 'the institution of marriage is under assault by organizations on the left seeking to redefine marriage fundamentally'.[8] Kobach's strategy gained a momentum that some initially thought would propel him to the White House (USA Today, 2004). Despite their anti-immigration platforms, candidates like Kobach were successful in galvanizing the state's Black and Hispanic churches, many declaring their support on religious, anti-abortion and anti-gay marriage grounds. This backlash language was used so commonly among politicians and the conservative movement that angry protestors took the state's tourism slogan, 'Kansas: As Big as You Think', and mass produced an alternative version commonly seen on bumper stickers that reads, 'Kansas: As Bigoted as You Think'.[9] Ironically, Kobach was defeated partially because the press revealed his political alliances with White Supremacist groups in the area. Voters, it seems, will tolerate – even endorse – homophobia and sexism but draw the line at overt racism (Zeskind, 2004).

8 See www.kobach-forcongress.com/Issues/marriage.htm on Kobach and the Federal Marriage Amendment.

9 Kansas ranks dead last among the 50 states as a tourist destination.

The state's history and continuing sensitivity to issues of overt racism highlights an important point about the conservative backlash: it places a high value on certain rules of etiquette. Adherence to these rules is vital to maintaining public support in conservative Kansas. For example, in the city of Topeka, voters provided an initial win in the 2005 primaries to the state capital's first openly gay candidate for City Council, rather than vote for the candidate's virulently homophobic opponent, the granddaughter of Fred Phelps (of 'God Hates Fags' infamy). During the same election, residents in Topeka rejected a Phelps-sponsored repeal of a city government-hiring ordinance that included sexuality in its anti-discrimination clause. This initial controversy made national headlines (Wilgoren, 2005). When the same gay candidate ran in the actual City Council election, however, she was overwhelmingly defeated.

In April 2005, Kansas became the 18th state to adopt a constitutional amendment banning gay marriage, with a majority 71% to 29% in favour. Adopting the most sweeping language available, the Kansas amendment defines marriage to be between one man and one woman and also declares that only these 'traditional unions' are entitled to the 'rights and incidents' of marriage. This provision explicitly prohibits the state from authorizing civil unions for both homosexual and heterosexual couples and challenges the legality of companies issuing health benefits to gay couples and partners as well as the status of wills, powers of attorney and custody arrangements (Hanna, 2005). There are also implications for family law. A judge in Ohio, for example, reduced a charge against a man accused of domestic violence against his female domestic partner, reasoning that Ohio's domestic violence law pertains to relationships that 'approximate the significance or effect of marriage' and the new gay marriage ban in Ohio denies any kind of legal recognition of unmarried couples. Other judges have ruled against this argument but the question of what is at stake with gay marriage bans persists (Rothschild, 2005: A1). The impact, it seems, is for gay advocates to anticipate and the courts to decide.

the new moral crusade: strange bedfellows and international sex trafficking

While the current political climate surrounding abortion and sexual morality is difficult, there is little confusion about who falls on what side of the aisle. By contrast, conservatives' latest foray into the area of human rights and the international sex trafficking debate has made several strange bedfellows. Since 2000, religious conservatives have mobilized the administration around the issue of international human rights. The top priorities for this group include persecution of Christians abroad, such as in the Sudan, and international sex trafficking. Similar to the issue of abortion, conservatives have framed the trafficking debate by appropriating the language of 'moral repugnance', describing sex trafficking as 'modern-day slavery' and positioning themselves as emancipators. Decrying trafficking as a 'special evil', Bush used the issue to rally his evangelical base during the 2004 campaign (Melzer, 2005). At the same time, prominent US feminists have allied with the administration and conservatives on the issue of international sex trafficking. In this final section, I examine conservatives' appropriation of the sex trafficking debate and feminists' subsequent alliances with some of the country's most conservative elements. These alliances prompt a number of questions, including whether or not a feminist position is self-evident in the current debate and the political and theoretical consequences of mainstream feminists' position and political alliances.

Kansas' conservatives have played a key role in the appropriation of the international trafficking debate. In 2000, conservatives and mainstream feminists formed an uncommon coalition to create the Trafficking Victims Protection Act (TVPA), which aims to monitor countries' work in combating trafficking. Senator Brownback of Kansas mentioned above as a prominent advocate of the Mexico City Policy, co-sponsored the TVPA with Democratic Senator Paul Wellstone, who died in a plane crash soon after the Act was passed. Initially created to address both global inequality and other forms of international trafficking outside of the sex trade, the Bush administration chose to focus on sex trafficking (Katayama, 2005). Conservatives emphasized the criminal penalties of the bill, stipulating that nations that fail to meet a set of 'anti-trafficking' standards face economic penalties or even sanctions imposed by the US. The Act's current supporters range from prominent American feminists such as Gloria Steinem, Catherine MacKinnon and the National Organization for Women to some of the most ardent religious conservatives in Washington DC including Richard Land, head of the ethics and religious liberty commission of the

Southern Baptist Convention and a primary supporter of gay marriage bans and anti-abortion initiatives (Miller, 2003).

Despite the TVPA, there is evidence to indicate that the commercial sex trade is on the rise globally (Melzer, 2005: 2). In many cases, the US is spurring on the international trend by targeting 'black listed' countries such as Cuba and North Korea, while ignoring those countries where the trafficking problem is more severe, such as India and Thailand, and minimizing the US's own role in trafficking. Similar to the administration's policies on international family planning and AIDS, the US policy to curb international trafficking does not focus on harm reduction. Funding preference is given to groups that forcibly remove women from prostitution (Crago, 2003). This means excluding organizations that have been working to make the sex trade safer for women from the funding scheme. Further muddling US policy toward the international commercial sex trade is recent evidence that suggests that fewer sex workers are victims of actual trafficking than economic migrants who enter the commercial industry due to a lack of viable alternatives (Shapiro, 2004). The TVPA offers no legal assistance to women who were voluntarily smuggled into the US to work as prostitutes, but they are included in the count of trafficked victims (Allan, 2005; Nathan, 2005). The administration's current approach, then, conflates trafficking with prostitution. The problem, of course, is that for conservatives, prostitutes do not make good 'victims'. The US's approach curbs some basic protections that local NGOs had been able to provide in the past, such as contraception, counselling about abortion and facilitating networks of support for women in the industry. At the same time, under the TVPA, victims of trafficking face a number of obstacles and can be further prosecuted. Taina Bien-Aime, from the US organization 'Equality Now', highlights that while the TVPA provides for provisions for special visas, in order to qualify the trafficked woman must cooperate in the prosecution of her trafficker or face the penalties afforded to all undocumented immigrants – prosecution and eventual deportation (Melzer, 2005). In their ideological tilt and emphasis on penalties and enforcement, the TVPA and the Mexico City Policy have much in common while their methods of enforcement penalize women in similar ways.

The current fervour around trafficking recalls US moral crusades of the past. In the early 20th century, for example, industrial expansion and increased immigration fuelled a similar effort to save 'defiled virgins'. (Melzer, 2005). A US led crusade against prostitution resulted in the 1910 passage of the White Slavery Traffic Act that banned transporting women across state lines for 'immoral purposes' (Kraut, 1996). Nearly a century later, both the media and US conservatives seem tantalized by the plight of 'the trafficked Third World woman', sold by her family into the commercial sex industry without her knowledge and against her will, whose only hope for dignity and salvation is US intervention. As Gayatri Spivak has described this strategic and imperialist use of feminism in other contexts, this monolithic representation of Third World women continues to recreate the often-repeated spectacle of righteous 'white men, seeking to save brown women from brown men' (Spivak, 1988: 101). In this imagery, the reality that most trafficking victims are not victims of sex trafficking but victims of forced and unpaid labour is largely ignored (Nathan, 2005).

Although Kansas is landlocked on all sides, state conservatives, having heard the rallying cry from their leader, Senator Brownback, mobilized around the issue of international sex trafficking in order to complement the federal initiative. The move to create a state law prohibiting trafficking was initiated by the most ardently conservative players in the state legislature, including politicians and advocacy groups with strong anti-immigration and anti-abortion commitments and generally hostile to initiatives to end violence against women, such as Attorney General Phil Kline and law professor Kris Kobach, discussed above. Conservatives' most obvious obstacle, however, was that no victims of trafficking were known to authorities. Conservative leaders, then, made a tactical and strategic move – they turned to domestic violence and sexual assault organizations in the state hoping they might find a victim willing to testify about her plight in public hearings while also providing the veneer of political cooperation. Although the state's domestic violence and sexual assault organizations had little documentation about trafficking victims within the state, they cooperated, helping draft the legislation and providing testimony that emphasized the possibility of sex trafficking in the state while downplaying those systems that fuel trafficking. During the public hearings and minimal media attention given to the passing of this state law, no feminist or liberal organization took public note of this odd and contradictory alliance.

feminists' critical responses

The current political climate gives rise to several questions familiar to the debates of feminism: What does such a strong conservative backlash mean for feminist political engagement and a coherent feminist politic in the long-term? And what does this mean for a transnational feminist coalition trying to challenge the rise of global conservatism while also trying to dislodge First World hegemony?

In the short-term, the rise of conservatism has created a more fragmented and ghettoized US feminist politic. As conservatives have captured the public imagination on issues of sexual morality, the US feminist movement has responded by siphoning itself off into institutionalized 'issue-based' groups, fighting very particular battles and sometimes working at cross-purposes. Thus, the pro-choice movement focuses its attention on saving *Roe v. Wade* in the courtroom while Congress and state legislatures cut away at poor women, Third World Women and adolescents' access to reproductive healthcare. At the same time, in their fight to protect the allocation of federal money to victim service programs, the anti-domestic violence and sexual assault movement has made strategic alliances with conservative leaders who are outspoken about their opposition to abortion and support for gay marriage bans. Although US feminist response largely reflects the conservative movement's success in creating divisions within left-leaning movements, these partnerships raise cause for concern.

Feminists' alliances with conservative leaders hostile to women in other areas are deeply suspect in the long-term. Rabbi David Saperstein, a veteran lobbyist in Washington D.C. for liberal causes, notes that there is 'the danger of legitmizing leaders, viewpoints and organizations who are deeply problematic to you in other contexts' (USA Today, 2005: A1). Mainstream feminists' alliance with anti-abortion conservatives on the issue of international sex trafficking and their acquiescence in the appropriation of a discourse of 'morality' in both the trafficking and abortion debates is evidence of this danger. Such alliances obfuscate and make apolitical those issues that explain the existence of an issue like trafficking, such as the inequality between the global North and South and the West's rapid and increased deployment of military bases that spur the commercial sex trade in those areas (Katayama, 2005). Although these relationships are considered by both conservatives and feminists to be strategic and short-term, they rhetorically cement the Right's appropriation of the language of 'moral rectitude' and work to fragment a transnational feminist politic. What arises is a feminist establishment that is well suited to represent the interests and needs of a very small group of women – namely white, middle-class, heterosexual, US women. The ways in which structures of power and regulatory policies directly affect differently positioned women is absent from the discussion.

Within this framework, the analytical force of the notion of 'differences', both between women and as an interrogation of identity, is all too easily replaced by a preoccupation with binary oppositions, in language and politics, which work to 'absorb alterity into the hegemonic and familiar' (Gunew and Yeatman, 1993: 13). That is, as US feminism's mainstream establishment acquiesces to the language of 'morality' to address abortion rights and makes strategic alliances with conservatives while remaining silent about gay marriage bans, its overall message is to ignore how differences shape women's political positioning and relations of power. And as this fragmentation boosts the position of some, it 'colonizes and appropriates the pluralities of the simultaneous location' of others (Mohanty, 1991:71). The inclusion of the notion of 'differences' has thus far humbled feminism's attempt to create a single, liberating agenda in the name of a 'global sisterhood'. Examining these 'differences' in relation to the Right's appropriation of the language of moral rectitude and its effects in places like Kansas and around the globe provides a basis for formulating a feminist politic that includes new models of cultural translation. To ignore the powerful effect of these differences, despite the recent steep rise of conservatism, risks a political exclusion that undermines feminism's long-term theoretical and practical work.

acknowledgements

For critical insight, essential editing assistance and invaluable conversations, deepest thanks to R. DeNies, C. Hemmings, K. Gurd and D. Hermreck. Any mistakes are my own.

author biography

A native of Wichita Kansas, Almas graduated with honors from the University of Kansas in 2002 and earned her MSc from the London School of Economics with Distinction in 2004. In 2003, she completed a Fulbright Fellowship at Hebrew University in Jerusalem. Almas Sayeed is currently a Congressional Hunger Center Research Fellow with the Economic Policy Team at the Center for American Progress, a progressive think tank in Washington D.C.

references

Allen, V. (2005) 'Senate Defies Bush on Foreign Family Planning Aid' *Reuters*, April 6. available online: http://www.commondreams.org/headlines 05/04 06-09.html.

Center for Reproductive Rights (2005) *Myths and Realities: Debunking USAID's Analysis of the Global Gag Rule* Center for Reproductive Rights, http://www.crlp.org/hill_int_ggr_myths.html.

Center for Reproductive Rights (2005) *Breaking the Silence: The Global Gag Rule's Impact on Unsafe Abortion*, New York, NY: Center for Reproductive Rights.

Crago, A-L. (2003) 'Unholy Collaboration' *Rabble News*, May 15.

Farney, D. (1995) 'Paranoia becomes an article of faith in a Kansas town' *Wall Street Journal*, August 17.

Frank, T. (2004) *What's the Matter with Kansas? How Conservatives Won the Heart of America*, New York, NY: Metropolitan Books.

Fried, M.G. (2002) 'Abortion in the United States: barriers to access' in Saelman, J. and Bhattacharjee, A. (2002) editors, *Policing the National Body: Sex, Race and Criminalization*, Cambridge: South End Press.

Gunew, S. and Yeatman, A. (1993) 'Introduction' in Sneja, G. and Yeatman, A. (1993) editors, *Feminism and the Politics of Difference*, Australia: Allen & Unwin Publishing.

Hanna, J. (2005) *Kansas Marriage Amendment Approved by Wide Margin*, Associated Press, April 6.

Jones, S. (2005) 'Kansas Seeking Abortion Records in Probe of Child Rape' *CNSNews.com*, February 25.

Katayama, L. (2005) 'Sex Trafficking: Zero Tolerance' *Mother Jones*, May 4.

Kraut, A. (1996) *Records of Immigration and Naturalization Service, Part 5: Prostitution and "White Slavery," 1902–1933*. Published by, LexisNexis Academic Library, available online at http://www.lexisnexis.com/academic/guides/immigration/ins/insa5.aspedit.

Kurtz, S. (2004) 'The 'Gay' Election: Bad Timing for Democrats' *The National Review*, February 10.

Lerner, S. (2004) 'Ashcroft Terrorizes New York Doctors' *The Village Voice*, March 15.

Melzer, E. (2005) 'Trafficking in politics: Bush's strong rhetoric on sex slavery masks policy failures' *In These Times*, March 14.

Miller, K. (2003) 'Southern Baptists Battle Gay Rights' *Tennessean.com*. www.tennessean.com, 5th November 2003.

Mohanty, C.T. and Martin, B. (2003) 'What's home got to do with it? in Mohanty C.T. (2003) editor, *Feminism Without Borders: Decolonizing Theory, Practicing Solidarity*, Durham and London: Duke University Press.

Nathan, D. (2005) 'Oversexed' *In The Nation*, August 29.

NBC News (2005) 'Meet the Press with Tim Russert. Transcript of interview with Howard Dean, Chairman of the Democratic Party *NBC News*. online at www.msnbc.masn.com.

Page, S. (2005) 'Christian Right's Alliances Bend Political Spectrum' *U.S.A. Today*, Washington D.C., June 14.

Palmer, E. (1982) 'Traditional Catholics Seek Their Eden in Kansas' *Kansas City Star*, April 18.

Pollit, K. (2004) 'Roe=Dred' *The Nation*, October 15.

Rothschild, S. (2005) 'Gay marriage ban poses unintended consequences; partner benefits, domestic abuse cases affected in other states' *Lawrence Journal World*, March 28.

Shapiro, N. (2004) 'The New Abolitionists' *Seattle Weekly*, August 25–31.

Spivak, G.C. (1988) 'Can the Subaltern Speak?' in Nelson, C. and Grossberg, L. (1988) editors, *Marxism and the Interpretation of Culture*, Basingstoke: Macmillan.

Staff Writer (2004) *Ashcroft Addresses Abortion Records Request*, Associated Press, CNN.com from New York City, February 13.

Staff Writer (2004) 'Moore Wins Comfortably; Brownback, Republicans Win' *USA Today*, Topeka, KS, November 3.

Supplement to Preliminary Summary of Legislation (2005) 'Legislature, Kansas Legislative Research Department 6 April 2005. Available online at http://skyways.lib.ks.us/ksleg/KLRD/Publications/2005_Preliminary_Summary_Suppl.pdf.

Sylvester, R. (2005) 'Kansas AG Seeks Identities of Abortion Recipients' *The Wichita Eagle*, February 24.

Wilgoren, J. (2005) 'Vote in Topeka Hangs on Gay Rights and Vitriolic Protester' *New York Times*, March 1.

Zeskind, L. (2004) 'Kris Kobach Loads Up with Anti-Immigration Ammo' *The Pitch*, September 23.

doi:10.1057/palgrave.fr.9400284

83 AIDS stigma, sexual moralities and the policing of women and youth in South Africa

Catherine Campbell, Yugi Nair and Sbongile Maimane

introduction

HIV/AIDS stigma is a key driver of the epidemic in sub-Saharan Africa, undermining HIV-prevention and the care and treatment of people with HIV/AIDS (Ogden and Nyblade, 2005). In this dialogue piece, we seek to examine how the vicious stigmatization of people with HIV/AIDS is intertwined with the social construction of sexual moralities – serving to perpetuate the social exclusion of women and young people in South Africa. We do this through highlighting our research into community responses to HIV/AIDS in two communities (one peri-urban and one rural) in Kwa-Zulu Natal province. We conclude by discussing the ongoing complexities of our efforts to implement these findings through our involvement in a community-led HIV/AIDS intervention in our rural research community.

Fassin (2002: 65) refers to HIV/AIDS as 'the embodiment of history', the transcription of social relations onto physical bodies. Those most seriously affected by HIV/AIDS stigma are often those who were already socially disadvantaged before they contracted HIV/AIDS (Parker and Aggleton, 2003). In unequal societies, interlocking symbolic, material and institutional dynamics ensure that inequalities are perpetuated (Bourdieu, 1984, 1986). Foucault highlights the way in which power hierarchies are policed not only by *overt* systems of power and government (inherent in laws and police forces for example) but also by *covert* systems of power, processes of internal 'psychological policing' thorough which people at all levels of the social hierarchy are socialized to voluntarily behave in ways that support the *status quo* (Foucault, 1977; Rose, 1991). Our work draws attention to the way in which stigmatization is one form of covert 'psychological policing', whereby those who breach power relations of gender and generation are disciplined and punished. The stigmatization of people with HIV/AIDS serves as a

(132–138) © 2006 Feminist Review. 0141-7789/06 $30 www.feminist-review.com

mechanism for highlighting the fatal consequences for young people or women who step out of line.

research findings

In our work (Campbell and Foulis *et al.*, 2005b; Campbell and Nair *et al.*, 2005a), we identify three inter-linked clusters of factors driving stigma. The first cluster involves the very real fear of infection in communities where around 40% of pregnant women are HIV positive, coupled with confused information about how HIV/AIDS is transmitted, and the lack of social spaces where people can discuss often strange and unfamiliar AIDS education messages and 'translate' them in ways that are appropriate to their own lives.

The second cluster involves a combination of poverty and poor HIV/AIDS management services. The stigmatization of people with AIDS tend to be most severe among poor families. The practical, emotional and financial stresses of caring for someone dying of AIDS may be the final straw for households already struggling to survive. This situation is exacerbated in conditions where there are few health or welfare services to support such households. For very real reasons people are terrified of catching the disease or having to care for someone with AIDS. Such hidden fears often manifest in an irrational hatred for those who already have the disease.

The third cluster of factors – most relevant to this *Special Issue* – involves the sexual nature of HIV/AIDS. Our research highlighted much evidence for the stigmatization of sex, and most particularly the stigmatization of the sexuality of women and young people – the two groups most heavily affected by HIV/AIDS in South Africa.

Many parents simply refuse to acknowledge the very possibility of youth sexuality, or that their own children are sexually active – even in the face of overwhelming evidence (e.g. teenage pregnancies, or obvious symptoms of AIDS) – and would certainly never discuss sex with their children. This adult refusal to face up to the reality of young people's sexual lives, and the concomitant failure to respect young people's rights to protect their sexual health is part and parcel of the wider social exclusion of young people in South African society: exclusion from respect and recognition (with youth commonly stereotyped as 'mad, bad and deviant'), from the political arena, from community decision-making processes and from paid work (Campbell *et al.*, 2004; Campbell and Foulis *et al.*, 2005a).

This denial of young people's sexual desire and relationships was particularly strong in relation to girls. Mothers refused to discuss sex in any way that resonated with their daughters' personal experiences of desire and relationships. They would darkly and indirectly allude to a link between sex, shame and danger, rather than accepting that their daughters were sexually active and teaching them how to protect themselves from sexually transmitted diseases and pregnancy.

This denial of young girls' sexuality is related to the wider 'demonization of women' that the epidemic has provoked in sub-Saharan Africa (Leclerc-Madlala, 2002: 2). Many of our informants spoke of the way in which 'the weakness of women' fuelled the epidemic, insisting that it was women rather than men who were responsible for promoting sexual morality (despite much evidence for women's relative lack of power in sexual relationships).

The denial of the sexual rights of young girls was mirrored in the lack of respect for adult women's sexual autonomy, particularly in our rural study community. Rural women repeatedly said they had no power to influence any aspect of their sexual relationship with husbands because *lobola* (bride-price) had been paid for them. If they suggested condoms or refused sex, their husbands could either beat them, and/or go to their parents' houses and demand that the parents pay a fine for their daughter's refusal to fulfil the bridal contract. More often than not, women's parents would berate them and tell them to obey their husbands.

A key factor sustaining the stigmatization of sexuality and HIV/AIDS was the link between sex, sin and immorality. The church was the main contributor of the symbolic ammunition that sustained this link. People often referred to HIV/AIDS as God's punishment for sin and evil, linking it to Biblical prophecies about the end of the world.

What are the underlying drivers of this web of representations of women, young people and sexual morality? The ability of adult men to police the behaviour of women and young people is a key pillar of patriarchal and adult-dominated social relations in South Africa. Men have considerably more privilege and power than women or youth, in both private and public life. A key aspect of this involves power to control their sexuality.

High levels of youth HIV/AIDS, especially among young women, shows very publicly that these groups are sexually active, despite male and adult attempts to control their behaviour. Stigmatizing people with HIV/AIDS is a way of punishing those who 'step out of line' and challenge the traditional control of adults and men.

Male control over women and young people is strongly supported by conservative traditional leaders, and by the church. Traditional authorities see ensuring 'acceptable' sexual behaviour in youth and women as a key aspect of their role. Here too, the existence of the AIDS epidemic shows up their lack of authority, and AIDS stigma is part of their conservative backlash against this situation. Old-fashioned practices such as virginity testing have been reinstated, for example, where girls who pass this test are rewarded with goods ranging from higher social status to virginity certificates (Scorgie, 2002).

Even well-intentioned leaders – those seeking to stem the epidemic – may contribute to stigmatization by expressing strong public opposition to sex before marriage, referred to as 'bad behaviour'. With this kind of opposition, people are frightened to admit to having a disease linked with prohibited sexual behaviour. Households are frightened to admit that one of their members has the disease.

The church teaches that sex should be conducted within a faithful marriage and that sex outside marriage is a sin. The HIV/AIDS epidemic highlights how many unmarried people are sexually active and so it also highlights the way in which many people ignore the church's teachings. This shows that the church has lost some of its moral authority and power. One of the strategies that some representatives of churches use to try to get back this lost moral authority, is to say that people with HIV/AIDS are guilty of sin and immorality and their behaviour may even lead to the end of the world.

HIV/AIDS management programmes as a platform to challenge oppressive gender relations?

In an academic paper (Campbell and Foulis *et al.*, 2005b), we argue for two lines of urgent action by HIV/AIDS activists, based on our stigma research findings:

- to work in AIDS-affected communities to promote greater critical thinking about the links between HIV/AIDS stigma and the marginalization of women and youth – using a Freirian model of resistance-oriented consciousness-raising (Freire, 1970); and
- to work hand in hand with social development programmes promoting women's and young people's social and political participation and increase their opportunities for economic empowerment.

While we still hold strongly to both points, only the second one has 'taken root' in our non-academic/activist work. This does not imply any reduction in the strength of our commitment to the first goal – rather it serves to highlight the complexities of seeking to apply academic research in real-world settings, for reasons set out below.

Since completing our research, we have been invited to participate in setting up a programme to strengthen local community responses to HIV/AIDS in our rural research area.[1] Here, our explicit aim has been to move away from the dominant model of HIV/AIDS management where actors or agencies from outside the community 'intervene' to assist communities. Such approaches tend to use strategies formulated by academics, experts or professionals based on their own conceptions of what needs to be done, and have had little success to date

1 By 'we' we refer to the authors of this paper, through our membership of the 'Community responses to HIV/AIDS' team at the Centre

(Campbell and Nair *et al.*, 2005b). In contrast, the aim of our programme is to develop strategies formulated by local people in line with their own understandings of what is possible in the context of their own daily lives and local realities.

for HIV/AIDS Networking (HIVAN), a university-based NGO with a commitment to building networks to facilitate HIV/AIDS management.

At the time of writing, women form the vast majority of local people who have volunteered to participate in the programme. Given this, coupled with the strong gender focus of our research findings, we would have expected to have a strong and explicit emphasis on 'challenging unequal gender relations' in this work.

No one in this community has explicitly challenged our analysis of the links between AIDS stigma and the oppression of women in the research dissemination workshops we have held to discuss the implications of our research for the proposed programme. Indeed, people have continued to offer even more evidence and personal anecdotes of the way in which men serve as obstacles to HIV-prevention and AIDS-care efforts.

However, local women appear to see little value in embarking on the types of women's consciousness raising activities familiar to western feminists, and which have played a key role in the empowerment of women in more affluent settings. While we – as academics influenced by western feminism – would have wanted to see unequal gender relations as a key issue to be tackled very directly and explicitly in the project, project participants from the programme community do not.

Discussions about possible project strategies have yielded many requests from women for skills building in the areas of leadership and poverty alleviation, with an explicit emphasis on the importance of targeting women to participate in such training. However, the manifestation of gender inequalities in the private sphere of inter-personal, family or household relationship is regarded as a given that women should 'work around' – rather than an as issue to be directly tackled in its own right.

In line with our determination not to impose our own agendas on the programme, we have been careful not to force our gender analysis onto community consultation workshops and project strategy discussions. In a chronically marginalized community where their life experiences foreground poverty and lack of political representation as the greatest obstacles to their health and well-being, women appear to see little potential gain from directly challenging men – who are only marginally better off than they are in the wider context of economic and political inequalities that structure the wider context of life in South Africa.

Much remains to be learned about the extent to which efforts to build women's roles and talents in public areas such as leadership and poverty alleviation serve to increase their confidence and empowerment relative to men in two HIV/AIDS-relevant arenas. The first unsolved problem involves women's relationships with men in the sexual arena, linked to their abilities to protect their sexual health.

The second involves the balance of responsibilities within the household (linked to the way in which the crippling burden of AIDS-care falls overwhelmingly on women). The project has only been running for six months at the time of writing, and its shape and outcome are still very fluid. Yet even in its short life it has already yielded many lessons about the gap between academic analysis and action in highly marginalized social settings.

author biography

Catherine Campbell is Professor in Social Psychology at the London School of Economics and Adjunct Professor at the Centre for HIV/AIDS Networking at the University of KwaZulu-Natal. She has published widely on issues relating to health, sexuality, social identities and HIV/AIDS, and is author of *Letting them die: why HIV/AIDS prevention programmes fail* (Oxford: James Currey; Bloomington: Indiana University Press).

Yugi Nair is Senior Researcher at the Centre for HIV/AIDS Networking at the University of KwaZulu-Natal, and has 20 years of experience in community and social development in South Africa. She is currently Co-Director of the 'Community Responses to HIV/AIDS Project'. Her recent publications include *Understanding and challenging stigma.* She is currently engaged in research into the role of partnerships in HIV/AIDS management.

Sbongile Maimane works at the Centre for HIV/AIDS Networking at the University of KwaZulu-Natal, where she co-ordinates fieldwork and community liaison on the 'Community Responses to HIV/AIDS Project'. She has co-authored a number of papers on various aspects of AIDS-care and HIV-prevention (especially amongst young people), including *Supporting youth: broadening the approach to HIV/AIDS prevention programmes.*

references

Bourdieu, P. (1984) *Distinction: A Social Critique of the Judgement of Taste*, Cambridge: Cambridge University Press.

Bourdieu, P. (1986) 'The forms of capital' in Richardson, J. (1986) editor, *Handbook of Theory and Research for the Sociology of Education*, New York: Greenwood, 241–248.

Campbell, C., Foulis, C.A., Maimane, S. and Sibiya, Z. (2004) *Supporting Youth: Broadening the Approach to HIV/AIDS Prevention Programmes*, Durban: HIVAN, http://www.lse.ac.uk/collections/socialPsychology/pdf/final_hivan_youth.pdf, Downloaded August 2005.

Campbell, C., Foulis, C., Maimane, S. and Sibiya, Z. (2005a) 'The impact of social environments on the effectiveness of youth HIV prevention: a South African case study' *AIDS-Care* Vol. 17, No. (4): 471–478.

Campbell, C., Foulis, C.A., Maimane, S. and Sibiya, Z. (2005b) ''I Have an Evil Child at My House': stigma and HIV/AIDS management in a South African community' *American Journal of Public Health* Vol. 95: 808–815.

Campbell, C., Nair, Y., Maimane, S. and Sibiya, Z. (2005a) *Understanding and Challenging HIV/AIDS Stigma*, Durban: HIVAN, http://www.lse.ac.uk/collections/socialPsychology/pdf/Understanding_Challenging_HIVAIDS_Stigma.pdf, Downloaded August 2005.

Campbell, C., Nair, Y., Maimane, S. and Sibiya, Z. (2005b) 'Building contexts that support effective community responses to HIV/AIDS: a South African case study. Submitted for publication.

Fassin, D. (2002) 'Embodied history: uniqueness and exemplarity of South African AIDS' *African Journal of AIDS Research* Vol. 1: 63–68.

Foucault, M. (1977) *Discipline and Punish: The Birth of the Prison*, London: Penguin.

Freire, P. (1970) *Education for Critical Consciousness*, New York: Vintage Books.

Leclerc-Madlala, S. (2002) 'On the virgin cleansing myth: gendered bodies, AIDS and ethnomedicine' *African Journal of AIDS Research* Vol. 1: 87–95.

Low-Beer, D. and Stoneburner, R. (2004) 'AIDS communications through social networks: catalyst for behaviour changes in Uganda' *African Journal of AIDS Research* Vol. 3, No. (1): 1–13.

Ogden, J. and Nyblade, L. (2005) *Common at its Core: HIV and AIDS Stigma Across Contexts*, Washington DC: ICRW, http://www.icrw.org/docs/2005_report_stigma_synthesis.pdf. Downloaded 1 May 05.

Parker, R. and Aggleton, P. (2003) 'HIV and AIDS-related stigma and discrimination: a conceptual framework and implications for action' *Social Science and Medicine* Vol. 57: 13–24.

Rose, N. (1991) *Governing the Soul: The Shaping of the Private Self*, London & New York: Routledge.

Scorgie, F. (2002) 'Virginity testing and the politics of sexual responsibility: implications for AIDS intervention' *African Studies* Vol. 61, No. (1): 55–75.

doi:10.1057/palgrave.fr.9400285

83 | stakes and states: sexual discourses from New Delhi

Jyoti Puri

report one: at the site of stranger violence

In the early hours of Sunday 8 May 2005, a young woman from the state of Meghalaya, studying at a Delhi University campus and working at a call centre, was sexually assaulted by four unknown assailants in the Dhaula Kuan area of the capital city. She was walking with a friend toward a nearby restaurant after 2:00 am when the four men abducted and assaulted her in a moving car. The friend informed the police immediately with a description and registration number of the car but there was no action to prevent the crimes in progress. The young woman was left close to where she was first abducted and, ironically, close to the Delhi police station for Crime Against Women (CAW) some two hours later. The following day, Delhi's leading English-language newspaper, *The Hindustan Times*, headlined: 'South Campus Student Gangraped', but provided few details. For several days, newspapers headlined and detailed the assault and its investigation and what emerged is a story of violence, a survivor's courage, and sloppy inflammatory reporting. Even as I write: just one positively identified arrest of the person considered to be the car driver and procurer; rumours that the other three assailants hide behind high-placed political connections; and unconfirmed reports that the young woman has returned to Shillong, Meghalaya.[1]

In one respect, it is understandable why this grievous crime was widely publicized. Delhi's moniker as the 'Crime Capital', can be recast as the 'Rape Capital'.[2] The National Crime Records Bureau's (NCRB) reports for 2003 note that the rate of crimes against women in the larger cities is 18.1 compared to the national level rate of 13.2.[3] By itself, Delhi accounted for 14.8% of the total crimes against women in India, a whopping high of 30.5% of the total rape cases and 35.0% of cases of kidnapping and abduction.[4] That this is a small percentage of the total sexual assault against girls and women is sobering for it is widely believed that no more than 10% of such crimes are reported, a position that was endorsed by a Deputy Commissioner of the Delhi police, who has written about sexual violence against women.[5] Perceptions of an insensitive police force, fears of reliving the violence through a hostile judicial process, trauma imposed by a culture that taints rape victims and dread of dishonouring the family are among the reasons that women will not

1 To contest the discourse on sexual assault, I deliberately draw upon information from news media, especially English-language newspapers, avoiding data gathered from the police. For all crimes reported, the police file a First Information Report (FIR), which details the information of the complainant, the nature of the complaint, where, when, and under what con-

(139–148) © 2006 Feminist Review. 0141-7789/06 $30 www.feminist-review.com

report sexual assaults in the vast number of cases.[6] This young woman's withstanding of sexual assault under threat of a gun and testifying to the police are laudable acts of courage. In fact, the *Times of India*, another leading English-language newspaper, spoke of the survivor's bravery in setting aside the trauma and her determination to bring the assailants to justice.[7] A senior police official described her as an 'intelligent, brave, and sharp girl', in the same report. The newspaper was less sympathetic to the police and sharply criticized their inability to mobilize immediately and to capture the assailants. The Delhi Police motto, 'With You, For You, Always', had fallen awfully short for this young woman.

Several anomalies arise in the intense media spotlight on this crime. Notwithstanding the series of 'copy-cat' crimes involving the abduction and sexual assault of women since the Dhuala Kuan rape, as elsewhere, attackers known to women commit the bulk of sexual assaults. The NCRB reports for 2003 confirm that in 87.0% cases of rape, the offenders were known to the women as neighbours or close relatives. The spectre of an attacker approaching a woman with nowhere to flee from an alleyway is not only misplaced but also misleading in its emphasis.[8] Yet, this framework shapes the reports in English-language newspapers and magazines. Curiously, the few details to be revealed in a crime of sexual assault, especially in a highly publicized one, are mishandled. The young woman's age, which state she is from, which college she attended, varied. Her age was variously reported as 19, 20 and 22 years; she was said to be from Manipur (a state), or Shillong (capital city of the state of Meghalaya), or in most cases from the Northeast states, of which there are seven, including Manipur and Meghalaya. The crime has come to be known as the Dhaula Kuan rape case.

report two: violent intimacies

In a city where 174 rapes were reported between January and April 2005, why does this particular crime garner such publicity?[9] Why is it *un*important to get the details right about this young woman? Is she merely a symbol of women's extreme insecurity in Delhi? Or perhaps a symbol of what happens to wayward women (from the Northeast)? How does this incident become casual teatime conversation? Sexual discourses crisscross a variety of cultural contexts providing clues and, sadly, intimate spaces are just as saturated as are public and state settings.

On 11 May 2005, several family members were gathered at a dear cousin's home, including two uncles, two aunts, a man in his 50s related to my cousin through marriage, my cousin, my mother, and me.

Multiple conversations intermingle over tea. My attention is caught abruptly as the man in his 50s exclaims about the rape of the young woman in Dhaula Kuan and laments how the two women could possibly be out so late in the night, 'Going out for tea at 2:00 a.m., *Bhai sahib* (brother)!' Much to my surprise, my cousin weighs

ditions the crime occurred, along with other relevant information. While this information is confidential, especially in the case of sexual assault, my status as a researcher allowed access to such reports.

2 As reported in a newsletter of Saheli (2003: 2), a feminist non-funded organization in New Delhi that mobilizes against violence toward women and other social minorities. According to the Saheli report, Delhi accounts for more than twice the number of reported crimes compared to the national average among metropolitan cities.

3 Please see: http://ncrb.nic.in/crime2003/cii-2003/CHAP5.pdf, consulted on 9 September 2005. Crimes against women tabulated by NCRB include: rape; kidnapping and abduction; dowry deaths and attempts at homicide for dowry; physical and mental torture; molestation; sexual harassment; and importation of girls under 21 years. NCRB also maintains crime statistics for gender-specific laws against trafficking, dowry, child marriage, indecent representation, and *sati*.

4 Ibid.

5 Personal communication, August 2005.

6 Many Muslim women survivors' courageous attempts to file police reports and tell about the brutal sexual violence endured by them and murdered kin in Gujarat in February–March 2002 are a notable exception to this trend.

7 *Times Of India*, Thursday 12 May 2005, p. 3.

8 For example, see the cover of *India Today*, a leading news magazine, published in the wake of this crime, 30 May 2005.

9 *Hindustan Times*, 11 May 2005, p. 3.

in with our parents' wisdom about not staying out late. I am puzzled, she who is like a sister, knows from first-hand experience that sexual harassment is unrelated to the time of the day. An aunt agrees with her and I interject that as women we do not have the luxury of coming home by 5:00 p.m. due to work, errands, jobs, and what's more, many women expect to be mobile as and when we like. However, stereotypes about young women from the Northeast prevail. The man notes that since they are from the Northeast and out at 2:00 a.m., they must be soliciting (sex). Objections to why two sex workers would report abduction and rape instead of negotiating sex and money fall on deaf ears, for they are from the Northeast and therefore seen as incapable of being raped; allegations of rape can only be false or the result of a failed transaction in this questionable framework.

Racist sexist stereotypes of young women from the seven Northeastern states, as Westernized and licentious in their clothing and sexual behaviour, steer the conversation to how women's and girls' clothes explain occurrences of rape. The man says, 'I don't know about these girls, *Bhai Sahib*, wearing what they do, showing their belly buttons'. This, from the man who amassed unusual wealth by exporting 'Western' clothes over the last 30 years. I can hold back no more, 'It doesn't matter what a girl is wearing, no one has the right to touch her without her permission'. This seems unrealistic to the rest. There is more to come.

An aunt reminds us that it was not always the case that women's clothing incited sexual assault. She cites depictions of Hindu goddesses that celebrate women in scant clothing, and laments Muslim (Mughal) invasions as the turning point. According to her, *Purdah* (invoked here as the metonym of women's degradation, the symbol of women's sexual objectification) was introduced at the time. My cousin agrees with the steady decline in the status of women because of an external, non-Hindu influx. We reach an impasse quickly.

Intersecting discourses of women's respectability, where they come from, whether they are considered rapable, and what accounts for the assault against women frames the discussion. They are not spoken in one voice or through one position. The discussion is superficially woven together by a focus on women, exceptional violent heterosexuality, and Muslims. That which remains unstated, the unnamed subject of the discussion, is its weft. Tacit in the discussion is the subject position of the normative heterosexual, middle/upper-class, Hindu male. What is said and what goes unsaid in a discussion on rape reveals marked and unmarked subjects. Women from the Northeast, women who dress 'inappropriately', Muslims who brought a culture of women's repression are named, marked. The issue becomes why women get raped, not why men rape and how to prevent it.[10] It is not stated that the name of the one person under arrest marks him as Hindu; the men evading arrest appear to be from the upper classes; the bulk of sexual assault is at the hands of men known to women. Consensus may have been lacking in the discussion but there was coherence to the normative subject position from which it proceeded.

10 This was the questionable framework that led the vice-principal, Virender Kumar, of Kirorimal College, Delhi University, to institute a code of dress for women

The private space of the family as a site where the violence against women is discursively and materially reproduced does not come as a surprise. The role of women in rationalizing violence is about how badly we want to see violence as an exceptional act that can be prevented if we accommodate sufficiently. If we don't work too late, if we don't dress provocatively, if we are not poor, if we are not promiscuous, if we don't... then we will remain protected. A study by Saheli (2003) of judgments on rape cases between 1950 and 1999 reported that 54% of the cases occurred between 6:00 a.m. and 7:00 p.m., in almost all cases women were dressed in 'traditional' attire, and 70% of those sexually assaulted were under the age of 16 years.[11] Insofar as complicity is shorn of its negative connotations and seen as 'folded togetherness' (Spivak, 2003), women's complicities are about reducing the scope of sexual violence from the structural/institutional to the individual isolated level. Perhaps anything more is too much to contend with.

students, especially from the Northeast states. His solution to sexual assault and harassment was that women students from the Northeastern states should dress modestly by wearing *salwaar kameez*. See http://www.hindustantimes.com/news/7242_1397573,00180-007.htm for the outrage resulting from this offensive policy.

11 Saheli newsletter (2003).

That sexual violence is a means through which to construct social minorities, establish social and national histories, requires that we build on the analytical and political contributions of feminists.[12] Unlike media and popular discourses, critical feminist writings have rightly refused to record 'rape against women' and 'communal violence' on separate registers. In her incisive essay written in the aftermath of the unspeakable devastation that was Gujarat in 2002, Tanika Sarkar (2002) speaks of how Muslim women's bodies became the site of almost inexhaustible violence and of the enactment of a sequence of revenge against Muslims by organized Hindu mobs. Our task is to unravel how this sexual violence becomes a node of producing multiple points of difference through and against one another. Our struggle is to split the arc that moves surely, painfully in the intimacy of the family from blaming the victim to blaming young women from the Northeast states to blaming Muslims for the fact that four men will abduct and assault a young woman. The work of Uma Chakravarti (1989), Angana Chatterji (2004), Zoya Hasan (1994) and Tanika Sarkar (2001) highlights that the arc connects discourses of gender, sexuality, region, race/ethnicity and religion, and constantly rearranges them to manage structural arrangements of power. How women across ethnic, caste, religious, and class groups experience sexual violence varies. However, what holds together the sexual assault of a young woman from Meghalaya with that of the putative respectability of upper-class Hindu women, with that of Muslim women sexually brutalized in Gujarat in 2002 are women's bodies as metonyms of sexual, communal, and national differences.

12 Considerable feminist work has furthered our understandings of the imbrications of gendered and sexual discourses with those of nation, religion, and caste/class in the context of India. Selected contributions include Bannerjee (2005), Butalia (2000), Chakravarti (1996, 1989), Chatterji (2004), Hasan and Menon (2004), Mankekar (1999), Sarkar (2002, 2001), Sinha (1995).

Sexual assault is aligned with disputed histories of Mughal invasions to reconstruct the originary and the extant nation as ideally Hindu. Sarkar (1995) and Sarkar and Butalia (1996) have addressed the role of Hindu women in inciting violence against social minorities and reconstructing the Hindu nation, their words saturating the mundane conversations of the Hindu upper-class urban family. Not only does the pre-Mughal nation become questionably rendered Hindu, a site where women roamed without regard for how they were dressed, the present nation is also rendered a site where the presence of Muslims is questionable. Discourses in the intimacy of the family are inspired not by outrage against sexual assault; in fact, rape becomes an active site of their assertion with the assistance of women of relative privilege. The blame for the rape is shifted to the victim and, insidiously, to groups seen as beyond the pale of national belonging.

report three: violent cultures of the state

More than 45 heads of police stations and police inspectors gathered for a training session on gender sensitization and gender justice on 11 July 2005 in New Delhi. Of these, two were women. Those gathered were mostly in their 40s and 50s, largely dressed in civilian clothes, and with at least a bachelor's level degree. We spoke in a mix of Hindi and English that is typical of a section of the middle-classes in Delhi. To highlight the role of culture and history in constructing gender as unequal difference, I led a discussion on the meaning of normative categories of woman and man. The discussion was lively and the issues of how gender categories exercise constraints on personhood, conduct, roles, occupations, and so on were thoughtfully received although mixed with deeply held scepticism about women's self-representations (clothes, behaviour, etc.).

What did not go well was the discussion relating to sexual and physical violence against same-sex sexualities and gender minorities. A discussion on gender sensitization and justice is incomplete if limited to a focus on women because of the significant violence experienced by same-sex sexualities at the hands of the police. Published reports of the significant physical and sexual violence targeted at same-sex sexualities and gender minorities were confirmed through my research as well.[13] Interviews with *Kinnars, Kothis*, and working-class gay men were filled with accounts of police violence throughout the research trip.[14]

I raised the matter of police violence against transgenders and same-sexualities. The audience overwhelmingly and unashamedly supported police action against *Kinnars*. 'They are criminals'; 'they rob and steal from their clients'; 'they are up to no good, it is our responsibility'; 'they do wrong; they solicit sex' were among the vehemently expressed sentiments. Where they were held back by beliefs or self-censorship about what could be said to sexual assault against women, no such constraints applied in the case of *Kinnars*. It did not seem to matter that *Kinnars* are heavily dependent on sex work, that they are vulnerable to sexual violence from thugs, goons *and* the police due to their extreme social marginalization. Rather, police (physical) violence against *Kinnars* was roundly defended in the interests of public morality and social order. And, this is where things became worse.

On the issue of same-sex sexualities, those gathered repeatedly and emphatically insisted that most crimes are committed by Muslims. Anal and

13 See the report by the People's Union for Civil Liberties, Karnataka on the violence against transgendered people in the Bangalore area. The report is available on the web: http://www.pucl.org/Topics/Gender/2004/transgender.htm.

14 *Kinnar*, also known widely *as Hijra*, are a long-standing, changing and hybrid gender and sexual identity within the sub-continent. Partly due to the derogatory use of the term and partly due to regional variations, the terms transgender,

oral sex, regardless of whether a same-sex or a differently-sexed couple is involved, regardless of whether consensual or forced, are crimes in India under Section 377 of the Indian Penal Code. The police are charged with the responsibility of enforcing this law, which is targeted toward transgenders and same-sex sexualities. The conditions under which the law is applied are too complex to elaborate here; suffice it to underscore the point that as a group they were resolute in their beliefs that most crimes under Section 377 and, indeed, most crimes in general are committed by those who are Muslim. One of the two women openly concurred. Non-Hindu members of the audience listened quietly to these vehement charges against a social minority group, perhaps cognizant of its explosiveness. That Muslims do not and could not account for the bulk of crimes in a country with more than 84% Hindus, that statistical evidence regarding Section 377 points away from Muslims toward Hindu names, fell by the wayside when gender was defined more broadly than women and issues of violence included those marginalized on the basis of gender and sexuality.

Kinnar (in the Delhi area), *Aravanis* (in the Chennai area) circulate along with *Hijra*. I alternate between transgender and *Kinnar* in this essay. *Kothis* are largely described as feminine-identified men from the working classes, who may or may not be married to women and who have sex with normatively gendered or hypermasculine men.

The state is another cultural node of the production and proliferation of sexual discourses. The discussions here serve as an important reminder that the state is not the material manifestation of the culturally imagined nation. Rather, the state is a messy set of institutions and their inter-relations that are as much characterized by threat and violence, as they are cultural sites marked by policies, ideologies, and myths, not the least of which is the enduring everlasting state. What lends weight to the sexual discourses that shape the above discussions is that they are expressed by those in state-sanctioned positions of authority. As police, these men and women mediate the privileges of citizenship and civil rights; they also withhold them legally, extra-legally and illegally. It is hard to predict the extent to which as police and as supervisors they draw a line between opinions expressed and policies followed. The muddying of the line between enforcing the letter of the law and policing as maintenance of public morality is troublingly evident in their responses to violence against gender and sexual minorities. Histories of police inaction and active complicity – politically-incited murderous rage against Sikhs in New Delhi in 1984 and more recently against Muslims in Gujarat in 2002 – are chilling reminders of the intricate conjoinings of discourses and practices characterizing the state.

report four: state discourses and policies

I attended a self-defence training class for nursing students at Ganga Ram Hospital, New Delhi, on 30 June 2005. This programme has been instated by the Joint Commissioner of Police (JCP), CAW Cell. According to her, since the highly publicized case of sexual assault on a nurse, it has become mandatory for nurses in New Delhi to undergo self-defence training. The nursing students, all women between the ages of 18 and 20 years, had completed approximately 25 hours of self-defence training led by two policewomen. They were lined and seated in six columns and nine rows in a medium-sized hall. Additional students and several faculty of the nursing school were in an adjoining area that was open to view.

For about an hour, the women exhibited what they had learned. They did a series of moves that could be used to stop, hurt, or incapacitate an assailant. They simulated how sexual harassment on the streets, in buses, spaces traversed routinely could be stopped easily. The young women showed how unusual cases, of being threatened with a knife, a pistol, of being choked, could be dealt with. Women students' accessories, a *duppatta*, a book, a pen, an umbrella were tools of self-defence. What was most impressive was the forceful projection of their voices as they demonstrated moves of self-defence. The voices resounded strongly, surely in the space.

The young women's enthusiasm was infectious. In our conversation at the end of the programme, they said they felt better prepared, less scared, confident that they would be able to deal with sexual violence. One woman said that she had already successfully stopped sexual harassment on a bus. The JCP's remarks that self-defence training helps unleash women's power and confidence was well reflected in the positive comments made by the student nurses.

The state is charged with protecting women and surely feminists have been at the forefront of enjoining its responsibilities not only to privileged women but also women at the margins of citizenship due to class, caste, race, religion, and sexuality. The CAW's programme to promote skills and the confidence young women need to protect themselves against sexual violence is an important step toward empowering women. In her remarks, the JCP underscored that womens' power is brought forth by the training. She reminded them that they might be physically smaller than their would-be assailants but they are not weaker; what they need to protect themselves they already have in the form of elbows, hands, legs and feet. According to the JCP, some 22,000 girls and women have received training in the past three years; the CAW cell cannot keep up with the demand.[15] The JCP encouraged the students to share their skills with friends and relatives so they, too, could protect themselves.

15 Personal communication, June 2005.

Even as CAW's training programmes are helpful and greatly needed, the emphasis on isolated stranger sexual violence leaves untouched the bulk of sexual assault against women perpetrated at the hands of a known person or organized through murderous mobs. Would it work to throw the violent neighbour over the shoulder? Probably. Would it work to jab a sexually aggressive boss with a pen? Probably not. Uncles, male cousins, neighbours, husbands, among those who are sexually violent toward young and older women, may use crude force but it is embedded in a complex series of emotional and psychological coercions against which physical self-defence is inadequate. The programme leaves intact the sexual assault of women that is enacted at the nexus of gender, racialized communities, and nation, with the active complicity of state agents and institutions. It leaves unexamined the conditions under which women are sexually assaulted by figures of state authority.[16] Those persons who are positioned at the margins of society because of their anormative gender or sexuality expression would not even qualify for the CAW programme. The problem is that the structural institutional patterns of sexual violence against women appear to have little influence on the CAW programme. Structures of heteronormativity, Hinduism, masculinity, dominant nationalism that shape expressions of sexual violence against women, transgenders, and same-sex sexualities remain intact.

16 The issue of 'custodial rape' has been especially charged in the context of India and carries a higher level of punishment of ten years and possibly life imprisonment. Dalit, tribal, and socio-economically disadvantaged girls and women have been especially vulnerable to custodial rape. That middle-class women are not immune became clear in a highly publicized case when a young woman was sexually assaulted in Mumbai by a policeman in April 2005; what has come to be known as the Marine Drive rape.

re-stating: by way of a conclusion

The four reports included here underscore the need to be vigilant about the discourse of sexual assault on women. Insofar as sexual discourses are tools of power dispersal, they produce truths about subjects, structures, and histories at institutional sites, the news media, the upper-class hegemonic family, and the state, in this case. Through the reports, I sought to draw attention to how the discourse of sexual assault traverses putative distinctions between the public and private, the civic and the political, and in so doing gains strength.

It would be a mistake to understand the discourse of sexual assault as coherent or singular, whether across or within the institutional sites of family, the media, and the state. As suggested through the reports, liberal attitudes of the news media toward the survivor of sexual assault collide with their racist undertones, which resonate with a reactionary history of India produced through the family, and which are at odds with defending the violence against *Kinnars*, and so on. Rather, what critical readings of the news media, mundane conversations in intimate spaces and discussions with state representatives suggest is that meaning making proceeds from the upper-class Hindu male subject position. Unmarked precisely as a factor of its privilege, this subject position drives what is said, what cannot be articulated about sexual assault against women, why and which women are raped, under what circumstances, as a result of which histories.

To draw attention to the role of the state was another aim of this essay. In the analytics of sexual discourses, I am struck by how frequently the focus remains on the cultural terrain of the nation while the state as a cultural/material set of institutions is often missing from our analyses. I wish to underscore the centrality of the state as a key site of sexual discourses and cultural policies. Self-defence training is an important step in militating against sexual assault and not to be dismissed lightly. However, the intervention does not go far enough. It leaves intact the structural conditions of violence against women and against transgenders, and the spectre of male stranger assault, rendered synonymous with marginal social class/caste and religious minorities, is further strengthened through the promulgation of such self-defence programmes.

Finally, I wish to end with an account of sexual violence against women within the family. During the summer of 2005, another story preoccupied the English-language media, and continues to unfold. The coverage focused on Imrana, a 28-year-old married Muslim woman and mother, who accused her father-in-law of sexually assaulting her. What precipitated the issue into public limelight were the decrees of two Muslim religious heads that Imrana could no longer be a lawful wife to her husband. The sexual violence that Imrana survived and brought forward to the local governing body has become a personal and political battleground. Imrana's story has been less about the sexual violence that occurs at the hands of family members, the dubious interpretations of some local religious leaders, and the means to ensure justice and reparation; instead, it has been overdetermined as a story about the difficult circumstances under which Muslim Indians must live in the Hinduized national state. Underlying the debates about the need for a 'common civil code' that would apply to all religious groups and the sanctity of Muslim Personal Law are the intersections of gender, religion, class, nation, and state that make Imrana's an embattled story. The violence is manifold and one hesitates to repeat and reproduce it. However, it would also not do to leave it unsaid in the context of this essay.

One strand of the many stories that are Imrana's takes us toward hope against the relentless nature of these sexual discourses. A rally was held in Mumbai on 23 July 2005 on behalf of Muslim women's rights and was organized by a coalition of women-centred and feminist groups. Mostly women protestors, some men, and children demanded that Muslim women's rights be protected and extended, that religious leaders not adjudicate women's lives, and injustice toward women be stopped. Representing the women in her neighbourhood, a young woman, wearing black pants, a black top and a black headscarf spoke forcefully against the restrictions in women's lives. 'No more', she said, 'we will not take men and mullahs telling us how to live, taking away the pleasures of our lives'. What complemented her words were the comments of a feminist activist that Imrana's is not just a Muslim women's issue. Similar stances have been taken by Hindu and Christian councils at the local level. It was an injunction to complement our concerns for those at the margins by unravelling that which remains unexamined because it is normative; the solidarities with Muslim women resisting the demands of family, religious leaders, and the state need to be supplemented by showing the faultlines of gendered and sexualized *Hindu* ideologies and practices. We would do well to not lose sight of this injunction.

acknowledgements

My gratitude to the editors for their feedback on this report. Thanks, as well, to Harleen Singh and Avinash Singh for a stimulating discussion related to this essay. The conversation with Nandini Manjrekar was similarly useful in clarifying my thoughts presented here.

author biography

Jyoti Puri writes and teaches in the areas of sexualities, states, nationalisms, and transnational feminist critiques. Her book, *Woman, Body, Desire in Post-colonial India: Narratives of Gender and Sexuality* (Routledge, 1999), addresses how constructs of gender and sexuality are shaped across national and transnational contexts. A recent book, *Encountering Nationalism* (Blackwell Publishers, 2004), examines questions of nationalism and the state from a feminist sociological perspective. She co-edited a special issue on transnational feminist sociology for the journal *Gender & Society* (April, 2005) and also serves as a deputy editor for the journal. A number of related articles and chapters are published in feminist journals and edited volumes on sexuality and gender. She is the recipient of fellowships and grants, including a Rockefeller Research Fellowship and a Fulbright Senior Research award. Her current research is on sexuality and the state. She is Associate Professor of Sociology at Simmons College, Boston, U.S.A.

references

Bannerjee, S. (2005) *Make Me a Man: Masculinity, Hinduism, and Nationalism in India*, Albany, NY: State University of New York Press.

Butalia, U. (2000) *The Other Side of Silence: Voices from the Partition of India*, Durham, NC: Duke University Press.

Chakravarti, U. (1989) 'Whatever happened to the Vedic Dasi? Orientalism, nationalism and a script for the past' in Sangari, K. and Vaid, S. (1989) editors, *Recasting Women: Essays in Colonial History*, New Delhi: Kali for Women.

Chakravarti, U. (1996) *Gender, Class and Nation: The Life and Times of Pandita Ramabai*, New Delhi: Kali for Women.

Chatterji, A. (2004) 'The biopolitics of Hindu nationalism: mournings' *Cultural Dynamics: Theory Cross-Cultures* Vol. 16, No. 2/3: 319–372.

Hasan, Z. editor (1994) *Forging Identities: Gender, Communities and the State*, New Delhi: Kali for Women.

Hasan, Z. and Menon, R. (2004) *Unequal Citizens: A Study of Muslim Women in India*, New Delhi: Oxford University Press.

Mankekar, P. (1999) *Screening Culture, Viewing Politics: An Ethnography of Television, Womanhood and Nation in Postcolonial India*, Durham, NC: Duke University Press.

Saheli (2003) 'Newsletter' September–December Issue, New Delhi: Saheli Women's Resource Centre.

Sarkar, T. (2001) *Hindu Wife, Hindu Nation: Community, Religion, and Cultural Nationalism*, New Delhi: Permanent Black.

Sarkar, T. (2002) 'Ethnic cleansing in Gujarat: an analysis of a few aspects' *Akhbár*, no. 3, July, http://www.indowindow.org/akhbar/article.php?article=99&category=2&issue.

Sarkar, T. editor (1995) *Women and the Hindu Right: A Collection of Essays*, New Delhi: Kali for Women.

Sarkar, T. and Butalia, U. editor (1996) *Women and Right-wing Movements: Indian Experiences*, London: Zed Books.

Sinha, M. (1995) *Colonial Masculinity: The 'Manly Englishman' and the 'Effeminate Bengali' in the Late Nineteenth Century*, Manchester, NY: Manchester University Press.

Spivak, G.C. (2003) Public Lecture, University of Hawai'i, Manoa Campus, 26 January.

doi:10.1057/palgrave.fr.9400286

83 | rewriting 'the paedophile': a feminist reading of *The Woodsman*

Carol-Ann Hooper and Ann Kaloski

We'd like you to imagine this scene. You're a woman who a few days ago picked up the latest guy at your workplace. He's a bit mysterious, but seems an OK person – he's already proved himself to be not racist, homophobic or sexist – and he's cute. The first time you had sex it was intense, but pleasurable. You sensed earlier that he was troubled by something, and referred half-jokingly to his 'dark secret'. Now you've just had sex again, are lighting your cigarette, and your lover is sleeping, but restlessly. When he wakes you ask: 'What did you do, Walter?' and this time he answers 'I molested little girls'.

You laugh nervously.

He responds 'Twelve years in prison is no joke' as he gets out of bed, starts dressing.

You are silenced. You gaze at him.

This is a key scene in the film *The Woodsman*. Walter discloses his past to his new girlfriend, Vicki, and she looks at him anew – he has just turned from an ordinary guy into a 'paedophile' in her eyes.

What might it mean to gaze at a paedophile? Of course, many of us have looked at an abuser, or averted our look, or tried to avert our look. It's a gaze that – like Vicki's – is not enclosed within a one-to-one relationship but is, rather, embroiled in cultural significance. Currently in the West perhaps only The Terrorist comes close to The Paedophile in being locked within a plethora of words and a poverty of meaning. Both are shadowy 'bogey-men' figures who remain unidentifiable within the community while being culturally identified by, in the case of The Contemporary UK Terrorist, a vaguely Asian appearance and a backpack and, for The Paedophile, a white, dishevelled and loosely creepy appearance. Both are demonized and dehumanized in popular discourse, and in the case of 'the paedophile' sometimes also in social work practice, with significant impacts not only on men who sexually

(149–155) © 2006 Feminist Review. 0141-7789/06 $30 www.feminist-review.com

abuse but also their partners and children. This short piece focuses on *The Woodsman,* a 2004 film directed by Nicole Kassell and based on a play of the same name by Steven Fechter, which, we argue, attempts to counter that demonization, while not in any sense obscuring the distress and harm that child sexual abuse can cause.

If you haven't seen the film *The Woodsman*, here's the plot as we understand it. The central character, Walter, is a convicted sex offender, returning to the community after 12 years in prison. The film portrays his attempts to negotiate a new 'normal' identity and to reintegrate himself within a network of relationships with work colleagues, a new girlfriend, his therapist and parole officer, and follows the challenges he faces both from the reactions of others to knowledge of his past and his own continued attraction to children. It is set in small town, working class, racially diverse US society, and in many ways is a very American tale of a man finding a way to 'hang on' through hard work, persistence – and the love of a good woman. However, we think the film offers more, and can be seen as an unusual and constructive feminist tale.

The Woodsman is significantly different from previous feminist films which represent violent men or women's responses to violent men. A common theme in these has been the killing of violent men by women – in, for instance, *Fried Green Tomatoes at the Whistlestop Café* (director: Jon Avnet, 1991), *Thelma and Louise* (director: Ridley Scott, 1991), *Monster* (director: Patty Jenkins, 2003) – represented as an understandable response to the women's own victimization. In *A Question of Silence* (director: Marleen Gorris, 1983), three women who meet in a boutique kill the male owner after he catches one of them shoplifting, and the rest of the film makes their actions comprehensible by exploring each woman's history of oppression and/or victimization by men. Where violent men are not killed, at the very least women escape/reject them and leave for example, *Once Were Warriors* (director: Lee Tamahori, 1995). These feminist films were to a large extent concerned to counter the silence and minimization that surrounded violence against women – as part of a movement seeking recognition of women's oppression they were hugely important. However, the reliance on violent responses from women and/or on separation alone offered a very limited vision for change – not many of us want to end up driving off a cliff (*Thelma & Louise*), being imprisoned (*Question of Silence*) or executed (*Monster*), or can rely on getting away with murder (*Fried Green Tomatoes*), even if we find killing justifiable in these circumstances, and many of us live in some kind of ongoing relationship with formerly abusive men, just because they are mostly not the stereotypic stranger but our brothers, uncles, family friends, fathers, co-parents, etc.

The Woodsman is of course made in a different context from most of these earlier films. The issues of violence against women and children are no longer silenced or unrecognized. On the contrary, child sexual abuse has become one of the major social problems of our time and attracts regular media interest. The dominant construction of the problem as one of 'stranger danger', 'the paedophile' as monster/pervert, still, however, obscures the issues of masculinity raised by feminist theory and research, and the evidence of much more widespread abusiveness within society. The popular hysteria whipped up against paedophiles, for example in the News of the World's 2000 naming and shaming campaign, triggered both vigilante action and calls for communities to have a 'right to know' when sex offenders were released (which have so far been resisted in this country, but not in many parts of the US) (Bell, 2002).

Feminist thinking has had considerable influence within social work, but a combination of influences has led here too to a demonization of violent men, especially men who sexually abuse children who are the one group commonly excluded from the fundamental social work value of respect for persons and regarded as 'inherently recidivist', incapable of change (Scourfield, 2003; Milner, 2004). However, while recidivism rates are high, and change is clearly difficult, there is also evidence that the isolation of child sex offenders on return to the community contributes to the problem. Circles of Support and Accountability, which involve members of the community volunteering to meet regularly with an offender to help them desist from further offending have had some success in Canada, and some have now been established in the UK (*The Guardian*, 13.12.00 and 18.11.04). Milner (2004) has recently argued that an approach to violent men based solely on challenging their behaviour may have been ineffective partly because it is based on definitions derived from early feminist work which have become fixed in stone, often in distorted form, which obscure the complexity and particularity of experience and behaviour, and that such an approach may oppress women further by reducing the choices available to them (women's only choice in a social work context is usually to separate from a violent man). She notes the lack of recognition in the now dominant feminist-informed approaches to violent men of the kind of men she sees in her own practice, who no longer express an unchallenged entitlement to use violence but are troubled by their own behaviour and seek help to change. Scourfield (forthcoming, 2006) also argues for the need to go beyond the polarized positions in which approaches to violent men *either* emphasize challenging them *or* recognizing their pain (with the former criticized as 'men-bashing' and the latter as 'letting them off the hook').

In contrast to these polarized narratives, *The Woodsman* opens up the process of multiple ways of looking, significantly within a popular cultural setting. It's a slow, claustrophobic film with lots of shots of railings, window curtains and blinds, isolated figures walking, thinking, eating, working. The predominant images that remain with us are of a grey, grainy, grimy environment, shot through with occasional images of a red ball rolling along the ground which, rather obviously, represents those aspects of Walter's desire that are for children and therefore dangerous. Yet, significantly for a tale that is, in some senses, about illicit desire, this is not a seductive film; the director doesn't exploit techniques of subtle persuasion to draw her audience into any specific perspective on Walter or on 'The Paedophile'. Instead, she uses two main devices: the first is to offer viewers a way of gazing at Walter from multiple perspectives, the second is to avoid realist representation of sexual abuse.

First, then, the multiple perspectives. One of the key gazes is of course Vicki's, the girlfriend we mentioned at the beginning. She walks out on Walter after his revelation, but returns, saying she sees something good in him. Later she tells him that she was 'poked about' by her three brothers as a child, siblings she says she loves, and who are now 'gentle hard-working men'. But the film depicts many other gazes, and the shifting between these denies the viewer any simplistic identification or thinking. There is Walter's brother-in law, a Latino, who is grateful to Walter for being the only member of his wife's white family who supported him when they married, and who approaches him in a playful, masculine, laddish way, but becomes aggressive when Walter asks him if he ever has sexual feelings for his young daughter. There are Walter's employer who sees him as a 'good worker' and his workmates, who have their own hard lives, and who are hostile when they find out about his past, but ultimately get on with their lives and don't bother with him. Among the workmates is Eve, who seeks out information about Walter and 'names and shames' him after unsuccessfully trying to pick him up. Walter's therapist relates to him as a person trying to change and capable of making good decisions, while his parole officer calls him a 'piece of shit', and sees the returning of sex offenders to the community as risky and regrettable, requiring 24/7 surveillance until they are caught re-offending and returned to prison. Walter's sister Annette appears only at the end of the film but we gather that she refuses to see Walter, although whether she is angry with him because he abused her when they were both children (he tells his therapist he just smelt her hair) or for other reasons is not clear. There is a young girl in a red coat, Robin, who sees Walter as a friendly if slightly scary adult, who seems to understand and almost even accept, unhappily, that he is grooming her, perhaps because she is regularly sexually abused by her father. And there is another known child abuser, who Walter names Candy, and who he watches from his apartment window and writes about in his journal, seeing the covert motivations in his friendliness to young boys when others apparently cannot. 'Candy' doesn't see Walter until Walter beats him up, towards the end of the film, perhaps becoming the woodsman of the Red Riding Hood story by 'saving' the child 'Candy' is with from the 'wolf', and in the process beating up/defeating/distancing himself from the 'paedophile' part of himself. We never see 'Candy's perspective; he is the unredeemed and still monstrous paedophile, ending the film with a broken jaw and unable to tell his story. At the centre of the film there is also of course Walter's own gaze on himself, initially seeing himself as he anticipates others seeing him, as outside the human compact, communing only with the birds on his windowsill, but gradually gaining some self-respect as he struggles to overcome temptation and (provisionally) succeeds, and immediately after doing so aligns himself with more respectable forms of violent masculinity by beating up Candy!

The second narrative strategy we'd like to mention which is used to disrupt easy responses is in the representation of sexual abuse. While films about sexual violence have often portrayed it explicitly in order to sensationalize, increasingly some do so for a different purpose, to represent it as reprehensible, as in *The Accused* which sought to shift public opinion on rape, or *Mysterious Skin*, where the sexual abuse of boys is graphically, and very skilfully, depicted on screen. In contrast, *The Woodsman's* director does not show Walter's offences, so as viewers we never gaze on that kind of 'realist' representation. Instead, a number of scenes position the viewer in relation to desire for children, in various ways which together unsettle any fixed boundary around 'abuse'. First, we have the camera gazing at pre-pubescent girls in a shopping mall, or on a bus, and the girls at times looking right back at the camera with self-assurance. Second, we look at Walter gazing at the young girls, as the camera attempts to capture his temptation. Third, there is the red ball of desire and danger, which appears at various crisis moments in the film. It's a child's toy. It can be picked up by Walter, rolled back for a game with the child, or ignored. Fourth, we have a sex scene with Vicki in which Walter appears to re-enact his sexual behaviour with children – Walter takes control and Vicki is manoevered onto his lap. As viewers, we see longing and puzzlement on both faces as Walter strokes Vicki's body, pushes her hands between their legs – and smells her hair. Fifth, the 'grooming'[1] scene in which Walter sits on a bench in the park and Robin, who he has met there before, sits down beside him. After a short exchange, he invites her to sit on his lap; she says no, then tells him of her father doing things she doesn't like, then offers to sit on his lap if he'd still like her to, and he declines. Like so much of the film there is nothing absolute, or noble, or stirring about this scene: Walter saves himself by refusing to abuse Robin, and although he appears to feel empathy for her, he doesn't 'save' her: she goes back home to her abusive father.

1 'Grooming' is the term used to refer to the process by which abusers befriend children and win their trust prior to abusing them.

Another aspect of this strategy is that very limited space is given to Walter's backstory: we never find out exactly what he did. Avoiding the details is a precarious strategy in some ways: it allows viewers to sympathize more easily with Walter but risks obscuring the seriousness of his offences. This lack of biographical specificity, however, avoids the fixing of a single truth about his past which defines him in the present.

To sum up, ways of thinking about men who sexually abuse children have been limited by current constructions of 'The Paedophile'. Feminists have done little to challenge the demonization of these men, and indeed feminist thinking is used to justify it in some contexts. We would argue though that *The Woodsman* offers that most feminist of visions, the possibility of change. Its methods – of multiple viewing positions, of complex and ambivalent stories in which the relationship between sexual desire and abuse is far from clearcut, and of deliberately generating a sense of the (flawed) humanity of all – not only represent but also open up space for more heterogeneous narratives of child sexual abuse. In the film, the community has to change a little to enable Walter to live in it, and he has to change a good deal to be allowed to do so. In reality of course many perpetrators of child sexual abuse deny responsibility for their actions and show little interest in change, and Walter is by no means the typical offender. But in an interconnected world, to define any group as beyond redemption in a way that is permanently fixed restricts the potential of others too. Adults with a history of child sexual abuse, for example, who often have complex and ambivalent feelings about their abusers, are not helped by the lack of alternatives to the popular demonized image of the pervert/monster.

The change Walter makes is halting, provisional, very unheroic. Some viewers who are otherwise positive about the film, have disliked its 'happy ending' (in which Walter leaves his flat to move in with Vicki). But while this may be unrealistic (this is fiction, after all!) it is not in our view either the hellish vision of a survivor of abuse hooking up with a(nother) perpetrator (while Vicki has been victimized in the past, she is not in any sense a victim in this relationship) nor a utopia in which all is resolved. Rather this ending seems very provisional, a moment in the ongoing lives of two people who are more than their experiences of abuse. As viewers, we can't project their stories beyond a few weeks, or even days – anything could happen. But it is a moment of hope, and without hope no change is possible. In seeing this more positively and as part of an alternative vision to previous feminist films, we are not of course questioning the value and legitimacy of anger, separation, or violent self-defence (nor at a fantasy level of the kind of retaliatory violence depicted in films such as *Kill Bill 1 (director: Quentin Tarantino, 2003) and 2* (director: Quentin Tarantino, 2004) as responses to victimization. There is also however a different kind of power represented in *The Woodsman*, with its more inclusive vision, which extends care and compassion even (alongside the necessary controls) to a man who has in the past sexually abused children.

acknowledgements

We are grateful to the organizers of the excellent Feminist and Women's Studies Association Conference on *Gendered Violence* (Aberdeen, Scotland, September 2005) for the chance to present an earlier version of this paper, and to the participants in our session for their very useful comments.

author biography

Ann Kaloski is a lecturer at the Centre for Women's Studies at the University of York. Her main research and teaching interests are in contemporary culture and fiction. She is co-founder of the small press Raw Nerve Books and chair of the York Lesbian Arts Festival committee. www.annkaloski.net

Carol-Ann Hooper is a lecturer in social policy at the University of York. She is author of *Mothers Surviving Child Sexual Abuse* (Routledge, 1992) and co-editor with Una McCluskey of *Psychodynamic Perspectives on Abuse* (Jessica Kingsley, 2000).

references

Bell, V. (2002) 'The vigilant(e) parent and the paedophile' *Feminist Theory*, Vol. 3, No. 1: 83–102.

Milner, J. (2004) 'From 'disappearing' to 'demonized': the effects on men and women of professional interventions based on challenging men who are violent' *Critical Social Policy*, Vol. 24: 79–101.

Scourfield, J. (2003) *Gender and Child Protection*, Basingstoke: Palgrave Macmillan.

Scourfield (forthcoming 2006) 'The challenge of engaging fathers in the child protection process' *Critical Social Policy*, Special Issue: Gender and Child Welfare.

The Woodsman: from a play by Steven Fechter. 2004. Film. Directed by Nicole Kassell, USA: Newmarket Films.

doi:10.1057/palgrave.fr.9400287

83 | book reviews

Sex in development: science, sexuality and morality in global perspective

Vincanne Adams and Stacy Leigh Pigg (editors); Duke University Press, Durham, 2005, ISBN 0-8223-3491-7 £15.95 (Pbk) 360p; ISBN 0-8223-3479-8 £64.00 (Hbk) 360p

In *Sex in Development: Science, Sexuality and Morality in Global Perspective*, Vincanne Adams and Stacy Leigh Pigg bring together a series of rich and detailed ethnographic studies carried out by anthropologists over the past 10 years in Asia, Africa and Europe. These studies examine how reproductive and sexual health programmes attempt to put into practice a global agenda reflecting the international standards and technologies of medical science. The book aims to demonstrate how modern sexual knowledge and health programmes claim to be value free by 'objectifying' sex and sexuality in universal biological terms, but that they are in practice based on implicit moral assumptions, codes and objectives that combine to attempt to create a universal 'normal' sexuality. The authors also examine the implications of these programmes for diverse local sexual practices, meanings and moralities. The locations represented in the book are subject to a range of international visions of sexuality that circulate in and through political, economic, religious and cultural agendas in the family, community and modernizing postcolonial and postsocialist nations.

The ethnographic studies are organized into three main sections. The first is concerned with examining how existing reproductive identities and gender norms, inherently tied to sexual practices, interact with scientific efforts to create a self-conscious, autonomous subject who identifies his or her subjectivity through a medicalized sexuality considered separate from other domains of social life. Michele Rivkin-Fish examines Russian doctors' resistance to American pharmaceutical representatives' attempts to teach them a 'value-free' way to offer contraceptives to female patients. In Greece, Heather Paxson highlights the moral conflicts women experience as the result of family planning programmes, and in Uganda, Shanti Parikh analyses how recent youth sex education initiatives have resulted in alienating adult relatives, leaving young people to explore sexual risks and pleasures in experimental and sometimes detrimental ways.

The second section is concerned with how ideas about what constitutes 'normal' sex are introduced, and how these normativities depend on health workers identifying local practices as 'backward' and deviant. In Tibet,

(156–180) © 2006 Feminist Review. 0141-7789/06 $30 www.feminist-review.com

Vincanne Adams studies the tension between Buddhist ideas of fertility and the population control programmes of the Chinese government. In Papua, Leslie Butt examines how Javanese state health workers render deviant the ways in which Dani women challenge existing regimes of power and assert control of their sexuality and choice of partner. Heather Dell's research focuses on the changing sexual practices and moralities expected of Indian prostitutes and middle class wives by men who are influenced by the influx of Western porn and mainstream videos.

The third section focuses on HIV/AIDS prevention initiatives that promote sexual liberalism, and the fallacies of the moral 'high ground' they claim. Vinh-Kim Nguyen's study in the Ivory Coast explores how poverty and local constructions of masculinity influence men's sexual practices, identity construction and responses to the messages of safe sex and sexual openness as strategies for empowerment. In the final study, Lawrence Cohen examines the role and impact of attempts by HIV/AIDS programmes in India to identify different local, male homosexual identity groups and how Indian men's identification with these groups, such as the *kothi*, connects them to wider political interests and conflicts, such as Muslim–Hindu identity politics.

Overall, this collection makes an important contribution to fledgling debates on sexuality and development in a global context. While sex and sexuality have been an increasingly important element of population and health development programmes, publications dedicated to detailed analyses of these experiences have been scarce. As the editors note, the main contributions to date have emerged from feminist theoretical and ethnographic research, historically focusing on exploring the cultural diversity and meanings of sexuality, gender, reproduction and kinship in different countries and ethnic groups. More recently, this research has moved beyond the study of heterosexuality to explore male same-sex sexualities, connecting with the fields of sexuality studies, queer theory and HIV/AIDS. In this volume, by introducing us to the role and effects of specific scientific health and population discourse, and by making the link with morality, the authors help to expose the values implicit in modern health discourse on sexuality that so often remain unrecognized and unquestioned.

Most notably, while the different contexts each tell a different story, a common feature of all is that development and modernization inherently depend upon and construct the norm of the heterosexual, monogamous, nuclear family that controls its reproduction and takes responsibility for the reduction of health risks. The studies also reveal the greater importance given to the morality and control of female sexuality and reproduction, and the unpredictable, diverse and often contradictory results of the global health agenda in different locations. They confirm that it is only in the field of HIV/AIDS that same-sex sexuality is now recognized as a legitimate topic for research and health initiatives, although only to date in relation to men, and by self-identified gay, male academic scholars and non-governmental programmes seeking alternatives to mainstream international and state-sponsored projects.

It is important to note that there is no attempt by the authors to close this book with recommendations for future research, policy-making or practice. This is a major challenge left open for those working in the field of health development, from local health educators to international agencies and health ministries, all under pressure to develop successful reproductive and sexual health models. The volume also leaves unexamined a range of experiences and contexts that have yet to be documented and analysed at an academic level. Experiences in Latin America, for example, include feminist and LGBT (lesbian, gay, bisexual and transgender) academic and activist research, advocacy and community-based initiatives in the field of sexual rights and citizenship. Moreover, to date there are very few studies of women's same-sex sexual histories, practices and identities, or of how women who do not conform to the heterosexual norm are affected by and respond to health and other development programmes. While we wait for more ethnographic research to emerge, it would be helpful for studies of this nature to at least address this issue at a theoretical level, with references to publications available to date (see, e.g., Rosenbloom, 1996; Thadani, 1996; Jolly, 2000; Hawley, 2001; Lind and Share, 2003; Robertson, 2005). In this way researchers could challenge current assumptions that only male same-sex sexual practices, subjectivities and experiences of discrimination exist and merit attention outside modern industrialized nations, especially among low-income sectors of the population that are the priority of international development research, policy and practice.

Carolyn H. Williams

references

Hawley, J.C. (2001) editor, *Postcolonial Queer: Theoretical Intersections*, New York: State University of New York Press.

Jolly, S. (2000) 'Queering development: exploring the links between same-sex sexualities, gender and development' *Gender and Development* Vol. 8, No. 1.

Lind, A. and Share, J. (2003) 'Queering development: institutionalized heterosexuality in development theory, practice and politics in Latin America', in Bhavnani, K.-K., Foran, J. and Kurian, P.A., editors. *Feminist Futures: Re-imagining Women, Culture and Development*, London: Zed Books.

Rosenbloom, R. (1996) editor, *Unspoken Rules: Sexual Orientation and Women's Human Rights*, London: Cassell.

Robertson, J. (2005) editor, *Same Sex Cultures and Sexualities: An Anthropological Reader*, Oxford: Blackwell Publishing.

Thadani, G. (1996) *Sakhiyani: Lesbian Desire in Ancient and Modern India*, London and New York: Cassell.

doi:10.1057/palgrave.fr.9400288

The lavender scare: the cold war persecution of gays and lesbians in the federal government

David K. Johnson, University of Chicago Press, Chicago, 2004,
ISBN 0-2264-0190-1 £13.00 (Pbk) 312p; ISBN 0-2264-0481-1 £21.00 (Hbk) 280p

Like all civil service employees working during the Eisenhower administration, Madeleine Tress – a 24-year-old business economist at the Department of Commerce in Washington, DC – was required to pass a security investigation as a condition for employment. At her position for only a few months, on that April day in 1958, Madeline was led into a room by two male interrogators who began the 'interview' by asking her a few mundane questions regarding her name, where she lived, and her date of birth. 'Miss Tress', one of the interrogators then retorted, 'the Commission has information that you are an admitted homosexual. What comment do you wish to make regarding this matter?' (quoted in Johnson, 2004: 148). Shocked, Madeleine froze and refused to answer the question. The men disclosed that they had reliable information that she had been seen frequenting a gay bar, the Redskins Lounge, and they named a number of her lesbian and gay male friends. One of the men then sneered, 'How do you like having sex with women? You've never had it good until you've had it from a man.' Tormented into silence, following the interrogation, she refused to sign a document admitting her alleged 'crime'. The next day, Madeleine Tress handed in her official resignation. By the late 1950s, literally thousands of women and men working in Washington, DC experienced similar inquisitorial grillings conducted under the guise of 'national security'.

There are moments in history when conditions come together to signal a seismic shift in the social and political geography. Three critical moments sparked an era of fear, suspicion, and repression leading to the interrogation of Madeleine Tress. The first occurred during the Truman administration in June 1947 when the US Senate Appropriations Committee warned Secretary of State Marshall that a concerted effort was being carried out 'to protect Communist personnel in high places' (p. 21), and that this subversive project involved 'the extensive employment in highly classified positions of admitted homosexuals who are historically known to be security risks' (p. 21). In their attempts to counter these alleged security lapses, the Committee attached the McCarran rider to an appropriations bill giving the Secretary of State authority to dismiss any employee at his 'absolute discretion' to promote public security. A second critical moment occurred three years later, in February 1950, when a relatively young and brash US Republican Senator from Wisconsin, Joseph McCarthy, provocatively claimed in a speech in Wheeling, West Virginia that 205 'card-carrying Communists' worked for the US State Department. In part as a response to McCarthy's allegations, the third moment transpired when Deputy Undersecretary of State, John Peurifoy, testified at a Senate appropriations committee meeting on 28th February 1950 denying, on the one hand, that his Department hired Communists, but, on the other hand, disclosing that a number of persons had been fired for being 'security risks', including 91 homosexuals. These disclosures set off a firestorm. Within one month, Republicans in the Congress ordered investigations looking into the extent of the 'homosexual problem' and the 'infiltration of sexual perverts' in government.

Drawing on a variety of primary sources – many only declassified and opened to the public in 2000 – plus oral histories of people who lived and worked in Washington, DC at the time, and newspaper accounts, magazines, novels, and films, David K. Johnson reconstructs an era of persecution and purges of lesbian, gay, bisexual, and transgendered people, as well as their heterosexual allies, in what the author has dubbed the 'Lavender Scare' during the height of the Cold War.

In the mind of many officials and in the public imagination, homosexuals and Communists in government posed similar threats. According to Johnson:

> Both groups were perceived as alien subcultures that recruited the psychologically maladjusted to join in immoral behavior that threatened the nation's survival. Many claimed the two groups were working together.
>
> (p. 38)

It is important to note that the Soviet government itself criminalized homosexuality under Joseph Stalin and blamed homosexuality on the West as a product of 'bourgeois decadence'. The US countered by blaming homosexuality on a Soviet Communist international 'godless conspiracy'. The so-called 'Red Scare' was said to have been saturated with lavender: the colour associated with homosexuality at the time. Some US government officials connected the Comintern (an international Communist organization) with what they termed the 'Homintern', which they saw as an international homosexual conspiracy linked with Communists. As Johnson makes clear, although gay and lesbian US citizens were never blackmailed into divulging classified state information and that connections between homosexuality and 'security risk' were groundless, mere allegations of homosexuality triggered congressional hearings, Presidential Executive Orders (for example, Eisenhower's Order 10450, which extended and enlarged the Harry Truman loyalty/security programme to exclude explicitly those who engage in 'sexual perversion' from obtaining government jobs), as well as executive agency security briefings.

A determination of the exact numbers of women and men harmed by the anti-homosexual inquisition cannot be known since detailed records were not kept and many individuals simply resigned before they were interrogated. Johnson estimates, however, that approximately 5,000 federal agency employees lost their jobs on suspicions of homosexuality during the 1950s through early 1960s.

Although there are some glaring omissions in Johnson's book, including substantive discussions of Roy Cohn (a top aid to Joseph McCarthy) and FBI chief J. Edgar Hoover's involvement in the Lavender Scare, plus overarching connections between the anti-Communist campaigns with pervasive anti-Semitism in the United States, Johnson makes a valuable contribution to our understanding of an often forgotten and overlooked facet of US history, and to the sociopolitical mechanics of scapegoating, persecution, and marginalization.

Warren J. Blumenfeld

doi:10.1057/palgrave.fr.9400297

Locked in the family cell: gender, sexuality and political agency in Irish national discourse

Kathryn A. Conrad, University of Wisconsin Press, Madison, WI, 2004,
ISBN 0-299-19650-X £31.50 (Hbk) 196p

The notion of the 'family cell', which is the subject matter of this book, is taken from Foucault's *The History of Sexuality* which identified this as the form, in 18th century Europe, that 'made it possible for the main elements of the deployment of sexuality (the feminine body, infantile precocity, the regulation of births, and to a lesser extent no doubt, the specification of the perverted) to develop along its two primary dimensions: the husband–wife axis and the parents–children axis' (p. 4). This book, while acknowledging the significance of the family cell to all nations, associates its particular significance in the Irish context with the patriarchal system of familial relations that emanated from the forces of Christianity, but also from a British colonialism that identified the answer to 'the Irish problem' as lying in the trope of marriage in which the maiden Hibernia would be rescued and restored to order by a patriarchal John Bull. The centrality of the heteronormative family is echoed in Irish nationalist discourse, which located the Irish man at the centre of family and nation. Conrad identifies a pattern whereby family members hide transgressions in Irish families from the eyes of potentially disciplinary institutions (p. 9). The family cell in Ireland regulated itself then, according to Conrad, by keeping to itself and thereby keeping its instability out of view. This self-preservation drive within the family, in Conrad's view, promotes the interests of both the nation-state and capitalism.

Alongside her thesis that the family cell functions to maintain the heteronormative national *status quo*, the argument is also advanced that family can act as a site of resistance when it enables self-preservation and self-reliance in the face of colonization. The main aim of this book, however, is to show how private sexual relations and the operation of the family cell are the foundations upon which the public sphere is built and reproduced. This raises important feminist questions about how to engage in a politically productive way with the reproduction of the gendered public/private divide. A further theme of the book is the centrality of information and informing as key sites of political intervention and resistance. Conrad's research points to the frequency with which gay men and sometimes lesbians are identified as potential informers and thus constructed as 'foreign bodies in the cell, pathological invaders threatening the very coherence of the cell' (p. 16). The only strategy for changing the *status quo*, according to Conrad, is a form of informing that 'cuts across public/private lines to destabilize a limited vision of the family cell as the only appropriate private model' so that the world might be imagined otherwise (ibid.). All of the propositions and challenges raised so far are examined in the book via reviews of critical political moments as well as a close reading of selected novels and films. Although the selection of novels and films works well with the thesis of the book, I would have been interested in Conrad's rationale for the selection of particular texts.

Drawing on the example of Roger Casement, who lost the support of Irish nationalists when diaries detailing his homosexual encounters were made public, Conrad points to how nationalist discourses construct homosexuality as a foreign threat. She develops this argument in a discussion of the 'queer-as-foreigner' in 20th century debates over the decriminalization of homosexuality in Britain, Northern Ireland and the Republic. Her theme of the homosexual as potential informer and agent of instability is further developed in relation to the treatment of lesbians in Irish feminist writing and in the controversy surrounding the Irish Lesbian and Gay Organization's (ILGO) attempts to march in New York St. Patrick's Day parade. In an interesting inversion of the construction of homosexuality as a private status, Conrad notes with irony the ways in which gay and lesbian demands to march in the New York St. Patrick's Day are rendered political while NORAID and Friends of Irish Freedom are allowed to march under banners without being identified as politically controversial. All of these illustrate, albeit in different ways, how the homosexual is positioned as personifying 'the breakdown first of the family cell and then of the borders of the nation-state' (p. 22). However, Conrad's discussion of the decriminalization of homosexuality in Northern Ireland and the Republic reveals a victory over the heterosexual definition of privacy but a failure to achieve public recognition of the homosexual 'as a member of a class of persons' ensuring that the public sphere remains a space in which only heterosexuality is performed (pp. 46 and 47).

Noting that feminism as a site of potential alternatives to entrapment in the family cell was more the focus of moral panic in the 1920s than lesbianism, Conrad proceeds to suggest that feminism can be seen to have (in a similar move to that of nationalists in relation to feminism) silenced lesbianism in the name of 'freedom'. Conrad supports this claim via her reading of Gifford Lewis's biography of Eve Gore-Booth and Esther Roper, June Levine's *Sisters*, Edna O'Brien's *The High Road* and Levine's *A Season of Weddings*. She identifies a range of moves in which lesbianism is explained away, rendered invisible via silence about how denigratory constructions of lesbians were used to undermine feminists, and by the location of lesbianism outside the realm of Irish identity, as somehow foreign. Improvements in the status of queers in Ireland in recent years are put down to the value of the 'pink pound', as niche consumption replaces political citizenship in a nation-state that increasingly defines itself through economic and consumer culture.

Conrad also traces the 'discursively obscured link between the 'private' choices of women and the 'public' interests of the nation-state' via a discussion of the abortion debates in Ireland in the 1980s (p. 71). Although questions of the reproduction of the nation-state and population control were central here, it was the morality or immorality of the individual reproductive choices of Irish women that were the focus of these debates. This discussion is expanded via Conrad's readings of Nell McCafferty's *A Woman to Blame: The Kerry Babies Case* and Margo Harkin's 1989 film *Hush-A-Bye Baby*. The explorations of real and fictional lives in these texts demonstrate how 'the suppression and circulation of information affect individual agency' (p. 99). For example, Conrad argues that Harkin's film 'forces the viewer to accept that every possible end to the pregnancy narrative, every 'choice' she might make, is shaped by political and religious discourse' (p. 98). Conrad also discusses the X Case, the 1992 Abortion Referendum, and Edna O'Brien's novel *Down by the River*, all of which reveal how the assumed sanctity of the family cell is enforced via the silences of young pregnant women. Information is manipulated and exploited to serve the ends of those who want to maintain a notion of Irish specificity based on the constructed 'purity' of the family cell. In Northern Ireland, Conrad notes that the cross-community Northern Ireland Women's Coalition has not taken a pro-choice stance as a party because the issue is considered divisive of women. The lack of debate about abortion in the north is, for Conrad, a reflection of a 'desire to remain 'fetal' – a political entity unwilling to acknowledge fully its relationship to a changing Great Britain, Ireland, and Europe' (p. 115).

Another theme of this book is how the flow of information between public and private spheres is closely monitored by state and resistant nationalism in struggles over territory and ideology. This discussion is developed via the argument that the future of Northern Ireland rests on numbers with a Catholic majority being predicted by 2050. In this sectarian situation, Conrad argues that women are expected 'to sacrifice political agency for the so-called greater good of the (unionist/loyalist or nationalist/republican) community' (p. 119). She points to the absence of any condemnation of the firebombing of a women's health and reproductive advice clinic on the Lisburn Road in 1999, which would have been forthcoming if it had a nationalist, unionist or church group attachment. Conrad interprets the silence of politicians as evidence 'that reproduction remains the unspoken substructure of sectarian politics' (p. 120). Indeed, the notion of 'parallel consent' enshrined in the Good Friday Agreement means that 'groups *and* individuals currently can have access to a politically recognized self only if they define themselves in sectarian terms' (p. 121). It is women and queers, Conrad argues, who trouble this static view of selfhood and draw attention to 'the politics of information and the politics of space in Northern Ireland' (ibid.). Anyone who exposes the instability of the heteronormative and patriarchal family cell, for example, the unfaithful wife, or the gay prisoner becomes 'an informer'. This theme is developed in the book via a discussion of *The Crying Game*, which is based on Frank O'Connor's short story 'Guests of the nation', and Stephen Birkett's 1999 novel *Ulster Alien*, which is a 'coming out story' set in Northern Ireland. In each case, Conrad argues that private affective bonds are presented as the only hope of challenging the state of affairs in the public sphere and then these bonds effectively undermine that hope. Moreover, in her reading of Fiona Barr's short story 'The Wall Reader', Conrad suggests that 'privacy is allowed only in so far as it reproduces politics as usual' (p. 134). Access to the public sphere is shaped by patriarchal, heteronormative, and sectarian assumptions so that the self that engages in the public sphere 'can do so only if it conforms to those assumptions' (p. 137).

Conrad's political intervention is to call for more informers. Starting from her position that information 'shapes literal space' she argues that informing 'breaks the taboos and allows dissident voices to enter and reinvigorate a stagnant public sphere' (p. 138). The reproduction of sectarian communities and national communities relies on the reproduction of the family cell. For Conrad, true sexual agency could come about through informing (which might be via pamphlets, protests, or cultural texts) and could destabilize family, nation, and sectarianism, thereby providing 'an opportunity for more people to contribute to a more flexible, less oppressive definition of 'self'' (pp. 137–138). The clearing of discursive space that would follow on from such informing would, in Conrad's view, enable a reimagining and renegotiation of geopolitical space and so a future with more possibilities for everyone.

This book brings a discussion of novels, film, and critical political events together to examine the workings of the family cell in the context of Ireland. I think this approach works well and enables the author to address the many different gestures by which the family cell is reproduced. A particular strength of the book is the way in which events north and south of the border are interwoven and discussed simultaneously, which also necessitated a discussion of the British context. This is a rare accomplishment in the context of academic work on

Ireland and demonstrates that it can be done well. Another achievement of the book is the successful interweaving of questions of sexuality, gender, and feminism with the workings of nationalism, sectarianism, and diasporic politics of identity. A concluding chapter would have brought the different strands together potentially strengthening the theoretical argument of the book. However, this book represents a unique intervention in debates about sexuality, feminism, and nationalism in Ireland and is an accessible, challenging and enjoyable read.

Breda Gray

doi:10.1057/palgrave.fr.9400289

Sex, gender and science

Myra J. Hird; Palgrave, Basingstoke, 2004, ISBN 1-4039-2177-6 £17.99 (Pbk) 208p; ISBN 1-4039-2176-8 £54.00 (Hbk) 208p

Myra Hird opens *Sex, Gender and Science* with an observation from teaching that anyone who teaches on feminist or gender theory will recognize immediately: the keenness with which students are willing to explore gender, and to accept its social construction, but the parallel reluctance to relinquish the biological 'facts' of the sexed body. Her book has two primary aims: firstly, to make the argument that Western understandings of 'sex' are '...based less upon an actual knowledge of sex 'differences' rooted in morphology than in a cultural discourse that emphasizes sex dichotomy rather than sex diversity' (p. 2); and secondly, to draw on 'new materialism' in order to contest those cultural assumptions. For Hird, the new materialism '...refers to a significant shift in the natural sciences that emphasizes openness and play within the living and non-living world, contesting paradigms which posited a changeable culture against a static nature' (p. 145). She identifies four central tenets to the new materialism – non-linearity and self-organization, contingency, variation and diversity – and explores these via literature from both the physical sciences and from the social sciences, making a powerful case for the role of 'science-literate analyses of matter' (p. 13) within feminist theory.

The scope of the book is wide-ranging, beginning with a review of feminist writing on science and the cultural construction of materiality. Hird then sets out the features and implications of the epistemic and political shifts of the Enlightenment that facilitated a transition from the 'one-sex' body to the understanding of two different, and complementary, sexes, and explores the ways in which the 'essence' of sexual difference – in bones, hormones and genes – has been written into scientific accounts of the sexed body. Chapter 4 revisits evolutionary theory to argue that, contrary to the arguments from sociobiology and neo-Darwinism, Darwin's original thesis posits that nature can be seen (and is increasingly being seen within the physical sciences) as a 'complex open system subject to emergent properties' (p. 13), rather than one of conformity and linearity. The latter half of the book takes up these principles of the new materialism and non-linearity by looking at the appropriation of sexual reproduction as representative of sexual difference, the silencing of the sexual diversity of living matter, and the intersexuality of the human body. The final chapter reiterates the book's core argument that it is a mistake on the part of feminists to reject science studies as necessarily reinscribing sexual difference in ways which are harmful to women, arguing instead that new materialism offers novel possibilities for both feminist theory and praxis through its articulation of the inabilities of culturally dominant discourses of sexual difference to account for the immense diversity of the natural world. This, she argues, creates not only the possibilities for challenging sexual difference, but more fundamentally, for destabilizing assumptions about the sex/gender binary itself.

The book's argument is rigorously made and is illustrated copiously with the kind of examples that you want to read out loud to those around you. Mushrooms with 28,000 sexes, regenerating livers and human mouths alive with hundreds of bacteria, amongst a host of other examples, line up to illustrate the 'superabundant diversity' (p. 13) of nature. Hird describes at entertaining length the contortions that primatologists have gone through in order to find explanations for homosexual behaviour that conform to heteronormative understandings of sexual activity as necessarily oriented towards reproduction. The attribution of normatively gendered behaviours to plants and animals is also well-documented, ranging across accounts of marriage and monogamy throughout the non-human natural world, displays of 'modesty' from the pistils (designated as female) of flowers, 'divorcing' birds, 'promiscuous' pollen, and 'spiteful' monkeys who 'selfishly' prevent other monkeys from reproducing. Other case studies are more sobering. In particular, the discussion of intersexuality in humans, and of the interventions that a rigid sexual dichotomy demands, illustrates clearly Hird's central argument that biology, and the appreciation of diversity, offers an essential means of challenging sex dimorphism, whose cultural production makes its material consequences no less real for those falling foul of that particular nexus of knowledge and power relations.

Sex, Gender and Science is a useful antidote to the impasse between biology and social constructionism that is apparent in the prevailing reluctance to relinquish the certainties of the sexed body and to engage directly with science studies. It is consistently argued and engagingly written, with a helpful glossary at the end of the book to facilitate the science-literacy it advocates. In the penultimate section of the final chapter, Hird proposes that rather than resist diversity, we should learn from bacteria, the diversity of whose accomplishments take over a page to list, declaring finally:

> 'So in the tired game of identity, I would choose neither goddess nor cyborg (Haraway 1991). I would rather be a bacterium'
>
> (p. 151)

By the end of the book, I couldn't have agreed more.

Karen Throsby

doi:10.1057/palgrave.fr.9400290

Some of us: Chinese women growing up in the Mao era

Xueping Zhong, Wang Zheng and Bai Di (editors); Rutgers University Press, New Brunswick, New Jersey, and London, 2001, 224p, ISBN 0-8135-2969-7 £15.50 (Pbk); ISBN 0-8135-2968-9 £40.50 (Hbk)

At last, Anglophone readers have an alternative to the stereotypes of suffering and persecution that dominate the numerous autobiographical accounts of the Mao years. Mao's rule involved crimes on a scale and of an order that have still not been properly accounted for. The experience of many who grew up under Mao's 'red banner', including those contributing to this volume, does not deny this, but it suggests a much more complex history than the accepted narratives of the 'victim' genre offer.

Some of Us is a collection of nine memoirs written by Chinese women, mostly from privileged urban families and now working as academics in the United States. Some were members of Mao's 'sent down youth', whose adolescent years were spent thousands of miles away from home in distant villages and state farms. Often left at home to fend for themselves during years when their parents were sent away to undergo political training in 'cadre schools' during the Cultural Revolution, they tell of the delights and difficulties of dealing with the independence that political circumstance gave them. The bare 'facts' of their adolescent experiences are not too different from many of those who have written the familiar mainstream accounts. Yet wanting to account for their past from their detached position in the States, they offer an appreciation of the powerful effects of the Mao era in shaping their cultural and personal identities. In their memories of their adolescent dreams, of their rebellions against their parents, and their own emerging and powerful sense of self formed through the opportunities and limitations that these years represented, the volume thus gives us fresh perspectives on how and why young people of the time 'bought into' Mao's project of social transformation.

Reflections on the changing discursive meanings of gender, and the intersection between these and the contributors' formation of self is another prominent theme in this collection. Many of the contributors first came across the notion of gender as a critical category when they arrived as students in the United States. Armed with the critical insights offered by feminist theory, they could then revisit their experience of the Cultural Revolution's rhetoric and practice of 'sexual equality' to reflect on the particular influence it had exerted on their own maturation as gendered subjects. A number of the contributors describe their disdain of the subject positions associated with the discursive terms used at the time for 'woman' and 'housewife'. They describe the self-confidence that young women could acquire through self-identification as gender-neutral revolutionary 'youth' (*qingnian*). The notion of 'youth' of the Mao years had few explicit gender connotations, and the youthful (*qingnian*) ideal of equality was infused with the political commitments which seemed to transcend gender difference. However, thinking back, Wang Zheng suggests that the 'gender-neutral' rhetoric of these years was by no means devoid of gender meanings. It did not signify a notion of 'liberation' along the lines of the second wave of feminism in the West, but it did, 'demonstrate the party's attempts to situate citizens in a new kind of social relationship, to....redirect their ethical duties from their kin to the party and the nation.' While many might condemn this as the social engineering project of a state bent on social domination, few, Wang Zheng suggests, have commented that 'the enforcement of this scheme disrupted conventional gender norms and created a new discursive space that allowed a cohort of young women to grow up without being always conscious of their gender' (p. 52).

Recognition of the importance of difference in conceptualizing social and gender practice emerges as a powerful, if implicit theme. It also emerges as a powerful inspiration to the feminist commitments shared by the contributors to this volume. Naihua Zhang recalls the close friendships with rural women established during her years working on a farm in the distant north-east. She describes her sense of betrayal of the dreams they shared when she returned to the city to enter higher education, leaving them to a future of economic and cultural constraints. Her acknowledgment of the tensions between her own privileged position and the disadvantanged status of the women with whom she worked established an understanding of gender difference that had a formative influence on her personal development as a feminist. Similarly, Xiaomei Chen, the daughter of 'luminaries of the theatre' describes the sense of empathy she developed for people from very different backgrounds she met when working in a state farm in the early 1970s. The youthful ideals of equality and justice she developed in those years found a new form in her later commitment to exploring the possibilities of critical feminist theory as an academic.

Many other reflections on the meanings of gender and generation in the formation of self arise in this volume. The shifting boundaries between constructions of the female and the domestic and public spheres; the effects on personal development of an enforced – though often welcome – independence during adolescence; intense and difficult relationships with mothers who represented models of the gender equality of the time but who were not around to care for their children. Participating in street drama troupes, leaving home to go on revolutionary 'long marches' alongside millions of other teenagers, or organizing experimental agricultural techniques in the cold north-east – through these different recollections, the contributors create a powerful sense of the autonomy they acquired as children and teenagers that informed their personal and professional choices in adult life. These are extraordinary women, who whatever our 'outsiders' interpretations of their revolutionary pasts, have moulded their experiences of growing up during an extraordinary time into their feminist takes on their own presents. All acknowledge the enriching insights into their pasts that their positions of detachment as academics in the US offer. Detachment across time and place may also explain the nostalgic tones of many of these accounts. But as the editors of *Some of Us: Chinese Women Growing Up in the Mao Era note*, experience and memory are 'sites of struggle for historical representations and experiences.' The issue is not to acknowledge experience as 'historical truth', but to recognize their historicity, and therefore importance in contributing, alongside others, to a comprehensive analysis of historical processes.

Harriet Evans

doi:10.1057/palgrave.fr.9400291

Behind closed doors: domestic violence in India

Rinki Bhattacharya (editor); Sage Publications, New Delhi/Thousand Oaks/London, 2004, 234p; ISBN 0-7619-3239-9 £14.99 (Pbk); ISBN 0-7619-3238-0 £29.99 (Hbk)

Domestic violence is the most serious violation of all basic rights that a woman suffers in her own home at the hands of members within her own family. The manifold problems associated with domestic violence have been systematically exposed by data and in-depth work has been undertaken by several people in the women's movement. The few studies available indicate that physical abuse of Indian women is quite high, ranging from 22 to 60 per cent of women surveyed (Mahajan, 1990; Rao, 1997). Most of the available information consists of qualitative studies of very small sample size. The only large-scale indicator of violence against women is the data relating to crimes against women published by the National Crimes Record Bureau, Ministry of Home Affairs, Government of India. The records of the bureau reveal a shocking 71.5 per cent increase in cases of torture and dowry deaths during the period from 1991 to 1995 and may reflect increased reporting of violence.

This collection of essays on domestic violence in India is enriched by contributions from eminent academicians and activists who have contributed to the women's movement in India in various ways. The editor, Rinki Bhattacharya is a writer and freelance journalist, having reported on women's issues in India and abroad for over two-and-a-half decades and is presently the Director of Bimal Roy Memorial and Film Society, Mumbai. In the introductory chapter, Rinki discusses the status of women in general and violence against women in particular. She describes various practices that are discriminatory towards women, ranging from forced marriages to 'honour killings' in India to Chinese custom of foot binding. The facts and figures she gives are mostly Western-centric and at best, of the Indian diaspora in the US, although there is plenty of data available on violence against women in India (for example see ICRW and CDPA Report, 2000).

In Chapter 1, Anwesha Arya elaborates the dichotomy within the Hindu religion of being 'worshipped, on the one hand, as the ultimate creative principle, and controlled on the other as threats to that very creation'. Anwesha traces through historical and archaeological evidence within the Hindu religion to understand the role of goddesses in forming the present status of women in Indian society. Anwesha ends her chapter with the analysis that 'if Indian women are able to find a route for being defined outside the sphere of marriage as individuals, not wives or mothers or daughters or sisters, there is hope'.

In the chapter on the State response to domestic violence, Sobha V Ghosh gives a detailed analysis of the changes in law around 'cruelty to wives'. Sobha details the provisions of the Domestic Violence Bill, which is presently before the Indian parliament and describes the women's movement's engagement with the State on issues of violence against women. The manipulation of the Bill, which was initially drafted by the women's movement to a form that was not acceptable by the movement and the struggles thereafter have been aptly discussed. Sobha then discusses the role of the family members, male as well as females, in perpetuating domestic violence on women. Referring to Agnes (1995), she states: 'The politics of the household cannot be seen simply as a struggle between oppressing men and victimized women....women themselves may be inserted into a hierarchy based on age, marital status and maternal status'.

Police attitude towards women is discussed in Kalindi Mazumdar's chapter, which elaborates from the experiences of student placements in police stations in a residential area in Mumbai. From the placement experiences, the chapter describes the attitudes of police towards women, what comprises a *good woman* and what comprises a *bad woman*.

The bulk of the book (pp. 67–190) comprises of narratives of 17 women interviewed by the editor during a span of seven years from 1984 to 1991. It is these narratives, which provide the richness to this book, describing various ways in which women are controlled and abused by their husband. These women from various different geographical locations, belonging to different castes, religions and from varied class and educational backgrounds, are similar in what they experience *behind closed doors* – violence. This collection of women's lives and their struggles from the abuser is an eye-opening account in their own words.

A useful part of the book is a glossary of terms that have been used in the book, which includes some terms that are used in common parlance in India and will be useful for a non-Indian reader.

Kaveri Sharma

Agnes, F. (1995) 'State, Gender and the rhetoric of law reform', Research Centre for Women's Studies, S.N.D.T. Women's University, Mumbai.

ICRW and CDPA Report (2000) 'Domestic violence in India: a summary report of four records studies', International Center for Research on Women and The Centre for Development and Population Activities, Washington DC.

Mahajan, A. (1990) 'Instigators of wife battering', in Sood, S. (1990) editor. *Violence Against Women*, Jaipur: Arihant Publishers.

National Crimes Record Bureau Reports (1991 and 1995) Ministry of Home Affairs, Government of India.

Rao, V. (1997) 'Wife-beating in rural south India: a qualitative and econometric analysis' *Social Science and Medicine* Vol. 44, No. 8: 1169–1180.

doi:10.1057/palgrave.fr.9400292

Women who live evil lives: gender, religion, and the politics of power in colonial Guatemala

Martha Few; University of Texas Press, Austin, 2002, 208p, ISBN 0-2927-2549-3 £15.50 (Pbk); ISBN 0-2927-2543-4 £39.00 (Hbk)

Few situates her book as a contribution to important themes in the historiography of colonial Latin America: understanding both colonialism and community formation as dynamic and contested processes, and specifically analyzing how women exercised authority and power within these. The majority of her source material comes from Inquisition trials, supplemented by other ecclesiastical and secular lawsuits, correspondence of colonial officials, city council records and religious tracts. While acknowledging the ways in which official Catholic doctrines and colonial power relations shaped testimonies taken by the Inquisition, she found that the lack of 'rigid lines of questioning' (p. 8) produced rich sources on local beliefs about the supernatural and revealed both the formation of social networks and conflicts within communities.

Few's specific case study is for the city and hinterland of the colonial capital of Santiago de Guatemala between 1650 and 1750, with the bulk of the Inquisition cases falling in the middle of this period. Chapter 2 introduces this particular historical context for the reader. According to Few, economic growth and an increasingly diverse population in the late 17th century initiated a period of social and cultural change that created opportunities for lower-class, often racially mixed women in the city to exert a degree of power through their reputations for sorcery and healing. Ecclesiastical and secular officials viewed with alarm the blurring of ethnic divisions and expansion of opportunities at the margins of the official economy, and associated the resulting disorder with women's activities. Strikingly, although the court testimonies refer to the activities of male sorcerers and healers, almost all of those prosecuted were women. Few could have explained more fully the degree to which this gender bias reflected the concerns of the Inquisition officials and/or the Guatemalans who filed accusations. It also would have been interesting to know more about the social profile of the accused women.

Chapter 3, the most intriguing in the opinion of this reviewer, focuses on bodies as sites of both colonial domination and women's exercise of supernatural powers. Women were widely believed to have an 'intimate knowledge of the human body through their roles in birth and death, washing the newborn and the dead, lactation, menstruation, and caring for the sick and elderly' (p. 53). Moreover, through their control of food preparation, women 'symbolically penetrated men's bodies' (p. 53) by mixing parts of their body (such as hair, nail clippings or fluids) into food and drink, especially chocolate beverages. Denunciations and testimonies demonstrate that men and women of all classes and ethnic groups believed they were potentially vulnerable to the power of female sorcerers. Sorcerers could both affect the bodies of their victims, often inverting them in gender terms, and transform their own bodies into animals, birds, or even light, an attribution Few identifies with anxieties about women's unstable and shifting roles in this period. She provides vivid examples from the Inquisition trials, including cases in which priests and sorcerers entered into a public competition for power over the bodies each tried to possess through the use of rituals.

Chapters 4 and 5 examine the healing and economic aspects of sorcery, respectively. Guatemalans in the colonial period consulted both male and female healers who mixed herbal and supernatural remedies to illnesses; the only specializations that Few identified were that midwives were always women, and that some indigenous men were regarded as specialists in blood healing. Chapter 5 considers sorcery as both a livelihood for accused witches and a resource for clients who wished to improve their fortunes; small shopkeepers, for example, were both the targets and purchasers of spells. At least as significant as these topics (healing and economic strategies) are the larger themes about dynamics within communities and the interventions of ecclesiastical and secular authorities that Few analyzes in both chapters. The networks of magical specialists and clients reveal the formation of multi-ethnic and even cross-class networks in urban Guatemala by the 17th century. Given the nature of her sources, Few sees these networks during moments of social conflict, when certain members of a community chose to denounce their rivals as witches. She highlights the power of talk or 'gossip' to both build and rend women's networks, perceptively identifying the double-edged nature of success as a healer or sorcerer: '... the crucial public aspects of women's activities that attracted more customers ... also invited the state into community relations' (p. 124).

The book will appeal to a wide audience. Its clarity of style and compelling tales from the Inquisition make it accessible to students. Scholars of gender will be interested by Few's broader analysis of how sorcery provides a window into changing social relations under colonialism, and specifically the nature of power wielded by women within urban and village communities.

Sarah C. Chambers

doi:10.1057/palgrave.fr.9400293

Doing women's studies: employment opportunities, personal impacts and social consequences

Gabriele Griffin (editor); Zed Books, London, 2005, 272p, ISBN 1 82277 500 6 £60.00 (Hbk); ISBN 1 84277 501 4 £19.95 (Pbk)

This book results from an EU-funded research project on the impact of women's studies training and employment in Europe. The project explored three key issues: differences in women's studies training across Europe; employment opportunities for women's studies graduates; and the impact of women's studies on the generation of change in the workplace. The book aims to further explore, and to answer, some of the questions raised in the original research. It uses the project's findings to develop the individual chapters, written by the project's research partners, that constitute the book.

This book is a useful addition to work which considers women's employment patterns throughout Europe; a Europe that, as the editor points out in her introduction, is rapidly changing. In 2004:

> ten countries, previously described as accession countries, joined (the EU)... significantly shifting the contours of that Union, geographically, politically, ideologically, economically and socio-culturally.
>
> (p. 1)

As the Union changes, so too do understandings of education and training within a globalized knowledge economy, with gendered, classed and racialized implications for learners. Although '(i)n recent years, the European Union has made access to continuing education one of the priorities in the fight against unemployment' it has done so in a climate of 'promotion of economic development and competition in a global market' (p. 56). Indeed, this may be why the book considers 'training' rather than 'learning', but I would have liked to have seen more discussion and analysis of this. Gabriele Griffin explains that 'training' was used because a significant proportion of women's studies education happens outside of the tertiary education sector (p. 11), but I am not convinced this is sufficient. The main purpose of lifelong learning in Europe appears to be that it leads to economic participation and greater opportunities in the employment market. In its discourse of individual autonomy, lifelong learning becomes the responsibility of individuals to prepare themselves to take their place in a working society. Such a learning agenda (re)constructs the structural inequalities of gender, class and other differences, where certain types of knowledge, skills and work are valued above others and any changes in discourse, policy and practice from 'learning' to 'training' are political. I would have liked to have known the impact of this – personal, social and economic – for students of women's studies.

I was left wondering about the title of the book, which seems to me to be more about employment opportunities for women in Europe than about women's studies or about personal impacts and social consequences. It may have been helpful to have seen the three key aspects of the book title (employment opportunities, personal impacts and social consequences) used as an organizing structure, to enable themes to be drawn together and developed. While they are of course not discrete areas, and are inter-related and in many ways interdependent, there is little or no discussion of this nor analysis of what such interdependence might mean with regard to gender as well as 'race' and social class. There is also the additional issue of age: particularly relevant to women studying women's studies, with its high proportion of 'mature' students. The organization of the book is not helped by the lack of a concluding chapter to pull together some of the key themes and to (re)turn to the three aspects of the book's title.

That is not to say that this is not a worthwhile book, full of extremely useful data, tables and figures, including employment opportunities for women in Europe (Chapter 1 by Nicky le Fauvre and Muriel Andriocci); equal opportunities through cross-European perspectives (Chapter 2 by Isabel Carrera Suarez and Laura Vinuela Suarez); and educational migration, mobility and gender (Chapter 6 by Borbala Juhasz *et al.*). However, although these are all highly relevant to women's studies students, I remain unconvinced of any special argument here for women's studies, and think that the book will have an appeal that moves beyond women's studies. This is important, with the decline of institutional women's studies through much of higher education.

I am not arguing here that women's studies should be written out of the picture: far from it. The book serves an important role in recording and re-telling something of the impact of women's studies on its graduates as well as on wider society. The book demonstrates the (different) values of many women's studies students when it comes to employment practices, including a desire to 'work differently' and a high commitment to social change:

> Their awareness and knowledge about multiple discrimination mechanisms that operate in the labour market enable them to encourage the widespread adoption of more egalitarian principles in their workplace.
>
> (p. 59)

An additional particularly interesting aspect of the book is that of comparative research, and the authors identify some key questions and issues for researchers working with cross-cultural differences, especially with regard to cross-cultural/national perspectives on gender, race, ethnicity and nationality. The chapters on research processes (Chapters 7 and 8, by Gabriele Griffin and Jalna Hanmer, respectively) explore the impact of multidisciplinarity; communication and translation; as well as comparability of databases, especially with regard to questions about race and ethnicity.

The book will then be of interest to researchers working on cross-cultural projects as well as to those working on the impact of gender differences on employment and training in Europe. Although I was at times disappointed with the lack of a thematic approach to this book, I am sure this is a text to which I shall refer in my own research as a source of information and as a valuable resource.

Sue Jackson

doi:10.1057/palgrave.fr.9400294

Against empire: feminism, racism and the West

Zillah Eisenstein; Spinifex Press, Melbourne, Zed Books, London and New York, 2004, 256p, ISBN 1 842777 394 £55.00 (Hbk); ISBN 1 842777 395 X £17.95 (Pbk)

Zillah Eisenstein's *Against Empire* is a sincere attempt from the West to think and see humanity beyond the boundaries of colour, language and power. Eisenstein tries to see 'everything/everywhere' through the politics of humanity by thinking subversively and seeing comparatively in this semi-autobiographical journey. To achieve this, she chooses a radically plural standpoint to see beyond the West; to respect humanity, freedom and democracies that have existed everywhere. For a feminist living 'elsewhere' – in the third world – it is a pleasurable and rare experience to read this Westerner, thinking in non-Western ways.

Eisenstein opposes America's empire building based on capitalist exploitation, racial discrimination and masculine militarization. Tracing the empire's past through slavery and racism, Eisenstein explains how the empire 'universalizes' and legitimizes exploitation and oppression through the language of terror and war. Her effort is to unmask the 'fictions' of the West by studying the Others of the West; the Muslim, the Black, the East and its colonial experiences. She seeks a notion of 'other than Western' universalism and an understanding of the contributions of non-Western, alternative democracies.

As Eisenstein rightly points out, historically, more than knowing themselves, the colonized are forced to know the colonizers (pp. 28–29). While being 'forced' by our universities to read/study many Western thinkers who brought enlightenment to the world, I have always wondered what was 'universal' about the ethics or reason endorsed by someone like Immanuel Kant, who justified racism and considered women as brainless. However, in the third world academia, even today, there is not much space for such critical and political knowing/learning.

Feminisms in the third world have struggled for decades to look beyond the dichotomies constructed by Western epistemologies, although not always successfully. Feminists from the third world have always wanted the privileged colonizer to know and see themselves more deeply and Others more humanly so that the universal could become more 'polyversal'. Eisenstein shares this search for a polyversal humanity, which is diverse, inclusive and locational, expressing a 'connectedness rooted in multiplicity' (p. 219). For her, feminisms all over the world are/shall be located in such a connectedness and unity.

In this respect, the politics of colour becomes significant for Eisenstein to explore, as she does in chapters two and three. As she points out, 'the silence about whiteness presents it as though it were not a color; and colors everything else by doing so' (p. 59). The origin of the term 'women/men of colour' has a historical significance of striving for a language of unity for the 'coloured'. But as a concept, 'women of colour' renders a meaning which shows the White as 'the colour' and others as Other colours or the White as non-coloured and the Other colours as coloured. Here language (as in the term 'women of colour') cannot escape from inventing or contributing to power hierarchies.

Human rights for Eisenstein are universal, 'humanly given to anyone who is human'. From this position she explains how the United States' talk on human rights is a type of rhetoric and how ideologies of liberal democracy and its propagators manipulate and control the human rights discourse. If Eisenstein is right in historicizing and contextualizing the human rights discourse, then the so-called alternative democracies in the third world also must have always known the meaning of this discourse and must have actively participated in it. However, the contemporary reality tends to depict 'human rights' as another new phenomenon, imported from the West. Hence it hasn't been easy for third world intellectuals/activists to contextualize it in the field or to make this term 'political' enough. This has been true historically about almost any concept imported from the West, whether 'development' or 'modernity'.

Attempts to trace the origins of exploitation and oppression are extremely relevant. However, it is also important for both the colonized and the colonizers to locate the multiplicities of oppression from within and beyond and from the nation to the Empire. And so, when we in India try to trace alternative democracies in our past, we not only read Gandhi (pp. 101–104), Tagore (pp. 106–108) or Sri Aurobindo (pp. 108–110), but also try to understand thinkers like Phule, Periyar or Ambedkar[1] for their contributions towards making India meaningfully democratic.

1 Jotirao Phule (1827–1890), E.V. Ramaswamy Nayker (popularly known as Periyar (1879–1973), Bhimrao Ambedkar (1891–1956).

In India, religious politics have done more harm than good to its unity. And that is why whenever Indian nationalism's imagined community stood for 'unity in diversity', India's poor, women, tribals and dalits fought for diversity – whether in Tamil Nadu, North East or Kashmir. For the white people in the West and for the upper caste in the East, a major contribution by the anti-racist/anti-casteist feminist politics will be, as Eisenstein illustrates, a humanitarianism that is polyversal. Quoting Toni Morrison, Eisenstein points out, 'Definitions belong to the definers, not the defined' (p. 188). The attempts by the West to define the Black or the Muslim belonged to the Westerners; it doesn't belong to the defined. And equally problematic has been the attempts of the East (living in the West or the non-West) to define the dalit or the subaltern.

If the polyversal world, as Eisenstein explains, wishes to have a 'view from everywhere', then feminists across the world will have to see and think more on many other fundamentalisms beyond the Muslim. Or to be more precise, Western feminists will have to tell the West to open their eyes towards Christian fundamentalism. Yet, how do third-world feminists confront the Empire? Eisenstein shows a very rarely visible passion and honesty to say that it is important for feminisms today more than ever to think polyversally and to see human bodies beyond colour and language. Yet, I was left wanting more specificity about exactly what such a polyversal account would entail and for whom. What her book contributes to is the hope that polyversal feminisms of the West and the East will see beyond boundaries created between empires and nations and the definers and the defined by capital and power.

M.S. Sreerekha

doi:10.1057/palgrave.fr.9400295

Feeding anorexia: gender and power at a treatment center

Helen Gremillion; Duke University Press, Durham and London, 2003, 280p, ISBN 0-8223-3133-0 (Hbk) £69.00; ISBN 0-8223-3120-9 (Pbk) £16.95

Feeding Anorexia is a refreshingly original and insightful account of how anorexia is (re)constituted. Through careful ethnographic investigation and astute critical analysis, Gremillion shows how the regime of a leading treatment centre for eating disorder sufferers in the west of the USA regenerates the very power relations that it identifies as part of anorexia's conditions of emergence. Gremillion's analysis marks a welcome shift away from much feminist social constructionist and poststructuralist theory about anorexia of the 1990s that tended to begin from a generalized model for understanding how power acts on (female) bodies, then apply that model to anorexia. By documenting how power relations are negotiated and regenerated in a specific context, Gremillion presents a sophisticated account of the effects of dominant discourses about anorexia and a sensitive description of how anorexia is experienced by sufferers and their parents. *Feeding Anorexia* is, as its author suggests, important 'groundwork for advocacy' (p. 23) and has far-reaching political implications. It is just what feminist literature on anorexia has been waiting for.

Drawing richly on her ethnographic material, Gremillion unsettles a core presumption of much feminist literature on anorexia: that psychiatric and 'cultural' explanations for anorexia are historiographically and epistemologically opposed. Indeed, not only are pathologized and socio-cultural discourses mutually constitutive, she argues, but it is precisely through psychiatric claims to objectivity that dominant 'cultural' ideas about individual and familial health and identity are thrown into relief (p. 32). As well as undermining the perceived binary relationship between 'nature' and 'culture' in arguments about anorexia's aetiology, Gremillion helpfully disrupts the frequent presumption that agency is discursively opposed to docility, or subversion to conformity. Through detailed descriptions of the relationships, practices, protocols, and decision-making procedures in the treatment centre, she demonstrates the ways in which apparent acts of resistance on the part of anorexic patients often have conservative effects, and seeming acts of compliance can challenge the prevailing order by exposing contradiction and conflict in the treatment programme. For example, she shows that for patients diagnosed with 'typical' anorexia, acts of direct resistance, such as in expressions of willfulness or anger, become enfolded back into dominant clinical interpretations of anorexia and are read positively as evidence of a nascent sense of identity deemed to be lacking in anorexics. Conversely, it is through exaggeration or hyperbolic compliance with social and clinical norms relating to self-control and fitness, she notes, that anorexic patients uncover instabilities in the ideology and practices of treatment and, consequently, highlight the gendered character and effects of discourses about the 'healthy' body.

The argument that contradictory or conflicting discourses collide in anorexia is not unusual in recent feminist theory about anorexia. But what makes Gremillion's analysis different is her attention to the actual, situated effects of such contradictions. In Chapter 1, she details how the treatment programme's precise management of patients' calorific intake and exercise accentuates the gendered cultural imperative that sees the body as a resource for self-actualization and the female body in particular as an encumbrance in this project. Because decisions about treatment are presented to patients as objective criteria for health, with conflict or disagreement among practitioners concealed, patients experience their own resistance as evidence of pathology. In this way, Gremillion argues, treatment practices generate precisely the senses of dependency and lack of identity that they set out to correct.

Continuing her analysis of the effects of treatment, Gremillion turns her attention, in Chapter 2, to structural family therapy. She details how a contemporary ideology of 'minimal mothering', infusing the treatment centre's approach, leaves mothers of anorexic patients in an impossible position in which they are both invited to blame themselves for being too 'overbearing' and yet, on their daughter's discharge, expected instinctively to supply the right level of love and support for continued recovery. Mothers are expected to withdraw from the treatment process, to facilitate separation and individuation, and yet exclusion generates the very desire for interdependency it seeks to redress. Further ironies abound in relation to the treatment team's conception of itself as a 'substitute family', which Gremillion explores in Chapter 3. The hierarchical gendered structure of the 'therapeutic' family reproduces many of the conditions identified as constitutive of anorexia in patients' 'natural' families. For example, female caretakers tend to monitor themselves critically for signs of 'over-involvement' with patients, while in practice a close attentiveness to patients is often generated in response to harsh treatment interventions on the part of male psychiatrists. In the process, these substitute fathers, like patients' 'natural' fathers, are absolved from taking responsibility for any part they might play in their daughters'/patients' distress.

Gremillion's analysis, in Chapter 4, of the ways in which dominant categories of individual and familial health and illness are constituted through ideologies of 'race' and class, is rare indeed in feminist literature on anorexia. She reveals how the profile of the 'typical' anorexic as white and middle class is regenerated, in circular fashion, through the marginalization and exclusion of 'borderline' patients who are often working class and/or black and who are cast as beyond the help of treatment because of their alleged inability to strike a 'healthy' balance between compliance and resistance. In their guise as objective descriptions, diagnostic categories like 'borderline' thus work to neutralize the racist and classist ideologies that structure them such that the responsibility for their effects is carried by the 'borderline' patients themselves.

Feeding anorexia closes with an epilogue which points to narrative therapy as an alternative set of possibilities for treating sufferers of anorexia, and one which is compatible with Gremillion's analysis of power. While this section provides a closing note of optimism, I wondered whether Gremillion's enthusiasm for narrative therapy left her, at times, slightly more forgiving of its inconsistencies than she had been for more mainstream treatment practices. For example, the use of language like 'anti-anorexic life story' (p. 207) seems in tension with Gremillion's argument that narrative therapy does not necessarily render anorexia 'separate' from the person suffering it or 'bad'. However, such tensions are not altogether out of place in the speculative space of an epilogue and are no doubt already proving constructive in Gremillion's current research on therapies informed by poststructuralist ideas about gender and power.

Jo Brain

doi:10.1057/palgrave.fr.9400296

ITALIAN FEMINISMS: IN THE MIDST OF PRECARITY:
Feminist Review Call for Papers

This special issue of *Feminist Review* intends to focus on issues and debates within Italian feminism(s) in relation to a newly emerging feminist scholarship, activism and creative practice. Precarity has, especially among young Italian feminists, become an organising principle for articulating conditions of work, life, politics and creativity. *Feminist Review* intends to engage critically and imaginatively with these debates and developments. The focus on precarity aims to offer an in-depth account of emerging negotiations within Italian feminisms. We are seeking innovative theoretical, creative and empirical articles for an international readership. We particularly invite contributions in relation to the following themes:

- Post-Coloniality
- Politics
- Labour
- Space
- Sexualities
- Creativity
- Migration
- Citizenship

Established for over 25 years *Feminist Review* is a UK based academic journal run by an editorial collective. Building on previous collaborations, for this special issue members of the collective are working in partnership with an editorial team in Italy.

The deadline for the submission of complete articles (in Italian or English) is **31st October 2006**.

All articles are subject to the anonymous international referee process. The decision to accept or reject will be made on the basis of this procedure. We strongly advise contributors to consult previous and past issues of the journal. http://www.feminist-review.com/

Submissions of articles must be emailed in Italian or English to j.m.andall@bath.ac.uk. Communication in English can be conducted with n.puwar@gold.ac.uk

Feminist Review: Call For Papers

Numero speciale dedicato a
Femminismi italiani: discorsi e pratiche intorno alla precarietà

Questo numero monografico della rivista *Feminist Review* intende concentrarsi sulle tematiche e sui dibattiti interni al femminismo italiano e relativi alla precarietà: pratiche culturali e politiche, scrittura accademica e creativa, elaborazioni teoriche costituiscono tutti materia di interesse per il presente numero. Negli anni recenti, soprattutto fra le giovani femministe, l'idea della precarietà è diventata un modo per esprimere condizioni di lavoro, vita, politica e creatività. Il numero monografico vuole dialogare in modo critico e immaginativo con i dibattiti, le pratiche e gli sviluppi intrecciati con la precarietà, categoria ed esperienza che permette inoltre di accedere alle elaborazioni più nuove ed originali dei femminismi contemporanei. Invitiamo pertanto a contribuire inviando saggi teoretici, creativi o legati a ricerche empiriche che siano di interesse per un pubblico internazionale. Vorremmo invitare dei contributi soprattutto sui seguenti temi:

- Politica
- Lavoro
- Spazio
- Sessualità
- Migrazioni
- Postcolonialismo
- Creatività
- Cittadinanza

Fondata più di 25 anni fa, la *Feminist Review* è una rivista accademica britannica gestita da un collettivo editoriale. Per questo numero, alcune donne del collettivo stanno collaborando con un gruppo redazionale in Italia.

La versione finale del saggio dovrà essere spedita (in italiano o in inglese) entro il 31 ottobre 2006.

Tutti i saggi accettati verranno sottoposti ad un processo anonimo di peer review. Consigliamo chi volesse contribuire di consultare i vecchi numeri della rivista all'indirizzo http://www.feminist-review.com/

Articoli in italiano o l'inglese devono essere sottoposti da e-mail a j.m.andall@bath.ac.uk. Comunicazione in inglesi può essere condotta con n.puwar@gold.ac.uk.

feminist review

Feminist Review is a peer-reviewed journal

www.palgrave-journals.com/fr

Feminist Review is published by Palgrave Macmillan. Feminist Review is published three times a year. It is edited by a Collective which is supported by an international group of Corresponding Editors.

Correspondence

Contributions, books for review and editorial correspondence should be sent to:

Sadhana Sutar
Feminist Review, c/o Women's Studies,
London North Campus, London Metropolitan University
166-220 Holloway Road, London N7 8DB, UK.
E-mail: feminist-review@londonmet.ac.uk

Notes for Contributors

Authors should submit their article electronically to: feminist-review@londonmet.ac.uk
If e-mailing is not possible 4 hard copies can be submitted but authors should note this may cause some delay in the reviewing process. Submissions must include an abstract (300 words), author(s) biographical notes, 6 keywords, a current address and an e-mail address.

Please note that contributions will not be submitted to the review process until all of this information has been received.

We assume that you will keep a copy of your work. Submission of work to Feminist Review will be taken to imply that it is original, unpublished work, which is not under consideration for publication elsewhere. All work is subject to a system of anonymous peer review. All work is refereed by at least two external (non-Collective) referees.

Our full Notes for Contributors can be seen at: www.palgrave-journals.com/fr/instructions.html

Publisher

All business correspondence and enquiries should be addressed to Feminist Review, The Journals Publisher, Palgrave, Brunel Road, Houndmills, Basingstoke, Hampshire RG21 6XS, UK.
Tel: +44 (0) 1256 329242. Fax: +44 (0) 1256 320109.

Feminist Review is online at:
www.palgrave-journals.com/fr
Visit the journal's home pages for details of the aims and scope, readership, instructions to authors and how to contact the Feminist Review Collective and the publishing staff. Use the website to search online tables of content and abstracts; register to receive the tables of contents by e-mail as soon as each issue is published; and order a subscription, reprints, a sample copy or individual articles.

Subscriptions - 2006 subscription rates

INSTITUTIONS

Combined (online & print): EU £235, Rest of world £235/$398; Online only: £224/$378

INDIVIDUALS

Combined (online & print): EU £38, Rest of World £38/$64 Online only: £36/$61

Subscriptions – outside the USA

Orders must be accompanied by remittance. Cheques should be made payable to Palgrave and sent to: Palgrave Subscription Department, Brunel Road Building, Houndmills, Basingstoke, Hampshire RG21 6XS, UK. Where appropriate, subscribers may take payment into UK Post Office Giro Account No: 519 2455. Full details must accompany the payment.

Subscriptions – USA

USA subscribers can call toll-free on: 1 800 747 3187. Please send check / money orders / credit cards details to: Palgrave Subscription Department, Palgrave Journals Subscriptions, 175 Fifth Avenue, New York NY 10010, USA.

Prices are set in UK Sterling. Dollar prices are converted from UK Sterling at the current exchange rate. Accordingly, your credit card rate may vary slightly from the dollar rate shown. To obtain the exact dollar rate shown, please remit by check.
All prices, subscriptions and details are subject to change without prior notification.

Advertisements

Enquiries concerning print and web advertisements should be addressed to: Andrew Patterson, Office 1, Sampson House, Woolpit, Bury St Edmonds, Suffolk, IP30 9QN. Tel: +44(0)1359 242375. Fax: +44(0)1359 242880. E-mail: patads@aol.com

Reprints and permissions

For reprints of any article in this journal or reproduction rights, contact David Bull (d.bull@palgrave.com) at the Publishers address above.

ISSN 0141-7789,
Online ISSN: 1466-4380

Typeset by Macmillan India Ltd. Bangalore-25, India
Printed and bound by Unwin Brothers, Woking, Surrey, Member of the Martins Printer Group